Hands-on Experience with Altera FPGA Development Boards

Jivan S. Parab · Rajendra S. Gad
G.M. Naik

Hands-on Experience with Altera FPGA Development Boards

Springer

Jivan S. Parab
Department of Electronics
Goa University
Taleigão
India

G.M. Naik
Department of Electronics
Goa University
Taleigão
India

Rajendra S. Gad
Department of Electronics
Goa University
Taleigão
India

ISBN 978-81-322-3767-9 ISBN 978-81-322-3769-3 (eBook)
https://doi.org/10.1007/978-81-322-3769-3

Library of Congress Control Number: 2017956335

Printed on acid-free paper

This Springer imprint is published by Springer Nature
The registered company is Springer (India) Pvt. Ltd.
The registered company address is: 7th Floor, Vijaya Building, 17 Barakhamba Road, New Delhi 110 001, India

Foreword

The traditional teacher-centered classroom teaching is transforming into the newer student-centered approach to learning. During this transition, the teachers and students need to go through familiarization and training in the new pedagogy. This book entitled "Hands-on Experience with Altera FPGA Development Boards" is an effort by the authors to meet this challenge. The technology space is ever expanding, and it is not possible to teach all of it in the classroom teaching curriculum. It is true that students now have access to vast resources at their fingertips. However, a book of this kind, developed based on the experience of the authors in teaching this to their students, is more suited since it has been improved based on the feedback from the students who have used it in its early form. The authors and their peer group in their department have put in extra efforts to make it student-friendly. This is a third book in the series of books brought out by the group, specifically on the "hands-on-approach" to skill development.

Embedded systems are all-pervading and offer limitless possibilities in the use of FPGAs in systems of diverse nature. This book offers an in-depth, yet practical, explanation of the various elements that make up the subject matter. Understanding the contents of this book does not require high level of prior preparation. The case studies on signal processing and control application are very important for a beginner to put a practical system to work. The students and researchers who wish to explore this area will find it highly useful, shortening their learning time and get them onboard quickly. Authors have extensive experience in this field. They are in academia and understand the needs of students. Also, they have strong connection with industries and thereby have a good grasp of the present status. They have worked themselves on these systems, and hence, the book has a greater authenticity.

I recommend this book for intermediate programmers, electronics, electrical, instrumentation engineers, or any individual who is strongly inclined to take up his or her career in embedded C programming. I am sure the reader will experience learning embedded programming by example and learning by doing. Last but not the least, this book will certainly be a value addition to the field of reconfigurable embedded programming platform.

Professor Raghurama
Director, BITS Pilani, Goa

Preface

Microprocessor and microcontrollers have revitalized the instrumentation world and now become ubiquitous. However, due to their niche role, when a particular microcontroller is discontinued, the entire product based on it has to be revamped, and the evolution of the technology means that the newer upgraded versions cannot be used in its place due to binary and socket incompatibility. Another issue which arises is of redundant hardware in microcontrollers posing a basic bottleneck in system optimization—many resources remain unutilized for routine applications.

In order to achieve portability, power efficiency, higher throughput, and less latency, the only alternative is to use the soft processor cores with FPGAs for small- and medium-scale production as they become more economic as compared to ASICs. Many vendors have come out with readymade cores such as NIOSII from Altera, Picoblaze and Microblaze from Xilinx. Building the system on FPGAs with these cores will not only facilitate earlier and easier market opportunities but will also give the advantage of using readymade full proof design alternatives, reducing the inconvenience of committing mistakes and debugging. The present book will explore the "know-how" for synthesizing chips for every embedded needs.

Methodologies in digital design have undergone tremendous changes over the past three decades. The use of FPGA and HDL for implementing digital logic has become widespread in the recent past, and use of FPGA in embedded systems is increasing almost day by day. A sign of the increasing importance of this area is that most of the technical institutes and engineering colleges have incorporated FPGA as the core subjects.

The domain of embedded systems is quite large and is centered around general-purpose processors and microcontrollers. The Altera FPGA forum receives numerous posts by newcomers to the technology asking questions on configuring FPGA, interfacing SRAM, building NIOS II system—this book is for those users as it essentially addresses most of these questions. The motivation behind writing this book was to ease out the difficulties faced by the students and researchers, so that they are not dependent on their supervisors to understand the field of reconfigurable embedded platform. To this end, it has many worked-out case studies in different areas of electronics like basic digital designs, sensors and measurement, biomedical instrumentation. It is intended for graduate, postgraduate, and research students from the electrical, electronics, computer and instrumentation engineering backgrounds as a ready reference during their work.

We promise potential readers that this book will reduce the steep learning curve and will help them quickly develop their embedded systems application in the shortest possible time frame. We recommend that the readers begin by reading through the summary paragraphs of each chapter, which will introduce each section and provide an overall picture of how the book is organized and how it will help them in creating their own design.

We would like to thank our student community and friends—their work in various industries helped identify the problems used in the case studies.

Though this book is intended for beginners in the area wherein the students aspire to learn skills building FPGA platform, a prerequisite knowledge in C/C++ and HDL will greatly help in understanding the complexities more easily. Since these two languages are now part of regular curriculum, we feel the students can directly start working on case studies.

Taleigão, India

Dr. Jivan S. Parab
Dr. Rajendra S. Gad
Prof. G.M. Naik

Contents

About the Authors

Dr. Jivan S. Parab is Assistant Professor in the Department of Electronics at Goa University, India. He completed his Ph.D. from the same university with the thesis titled "Development of Novel Embedded DSP Architecture for Non-Invasive Glucose Analysis." He received his M.Sc. (2005) and B.Sc. (2003) in Electronics from Goa University. He has co-authored two books, published by Springer. The details of the books are "Practical aspects of embedded system design using microcontrollers" and "Exploring C for Microcontrollers: A hands on Approach." He has published several papers in national and international level journals and conferences.

Dr. Rajendra S. Gad is Associate Professor in the Department of Electronics at Goa University. He received B.Sc. (Physics) and M.Sc. (Electronics) degrees from Goa University in 1995 and 1997, respectively. He completed his Ph.D. in Electronics in 2009 from the same institute. He has several papers published in journal and conference proceedings to his credit. His areas of interest are biomedical sensors, DSP digital repositories and networks. He has been into teaching and taught courses such as VLSI system design, HDL system design, digital signal processing, computer programming, operating system, mechatronics, and electronics practical.

Dr. G.M. Naik is Professor and Head of Department of Electrics at Goa University. Dr. Naik's areas of interest are fiber optics and sensors, opto-electronics, renewable energy sources, and biomedical instrumentation. He completed his Ph.D. (Opto-electronics) from Indian Institute of Science, Bangalore, in 1987. He received B.Sc. (Physics, Chemistry, and Maths) and M.Sc. (Applied Electronics) degrees from Karnatak University in 1978 and 1980, respectively. Dr. Naik has co-authored two books entitled "Practical aspects of embedded system design using microcontrollers" and "Exploring C for Microcontrollers: A hands on Approach" published by Springer. He has several papers published in journal and conference proceedings.

Chapter 1
Genesis of PLD's, Market Players, and Tools

Contents

Abstract "Genesis of PLD's, market players, and tools" discuss the microprocessor, microcontroller, and PLD devices and also talk about how to select these devices for desired application. This chapter gives the family tree of PLD devices and helps designer to select best PLD devices based on application. The chapter also gives the overview of major PLD market players and programming aspect of VHDl, Verilog, and ABEL. There are several separate books available in the market which discusses in detail about VHDL, Verilog, and ABEL programming. Here, we simply focused more on the basic part of hardware descriptive programming language.

Keywords PLD · VHDL · Verilog · ABEL

J.S. Parab et al., *Hands-on Experience with Altera FPGA Development Boards*,
https://doi.org/10.1007/978-81-322-3769-3_1

1.1 Brief Insight of Microprocessor, Microcontroller and PLD's

Microprocessor

Microprocessor in any embedded system design is like a human brain, which provides computational control and decision-making capabilities. Microprocessors find use in advanced electronic design systems such as printers, automobiles, defense. In general, microprocessors have ALU, control logic to generate various control signals, and registers to store data required for processing unit.

Classification of Microprocessors

Microprocessor classification is based on function handling and features supported. The several companies manufacture many variants of microprocessors currently available in market but most frequently used microprocessors are as follows:

Intel microprocessors

4-bit processors: 4004, 4040
8-bit processors: 8008, 8080, 8085
16-bit processor: 8086, 8088, 80186, 80188, 80286
32-bit Processor: 80386, 80486,
64-bit processor: Itanium, Dual core, i3, i5, etc.

Zilog microprocessor:

8 bit processor: Z80, Z180

Motorola Microprocessor:

8 bit processor: 6800

PLD's

A programmable logic device (PLD) is the device in which the designed logic is implemented and easily reconfigured by the programmer on the fly. These devices are called as field programmable logic devices since the designer has flexibility of device programming in same field. The PLD gives designers the flexibility to implement different complex designs for various applications. Programmable read-only memory (PROM) is the most commonly used PLDs. There are two categories of devices: (a) devices are programmed by the vendor using a mask, and interconnects are one-time programmable, (b) devices that are programmed by the user are called field programmable. PLDs are very much inexpensive and flexible which are the biggest advantages.

1.1.1 Selection of Technology Based on Application

In embedded system design, the processor plays an important role on the designed system's success or failure.

Selection of the proper device for right application is therefore extremely important. Embedded application devices are broadly divided into microcontrollers and microprocessors. MPUs come in an extensive range of different types, models, and sizes.

Choosing between a microprocessor, microcontroller, or PLD's is a complex and rather daunting task. Several device selection criteria are discussed below. Selecting the proper device on which to base your new design can be daunting. The need to make the correct balance of price, performance, and power consumption has many implications.

Processing Power

The initial selection criterion is performance; microprocessor unit (MPU) offers more processing power than microcontroller unit (MCU). A broad comparison between devices can be made by comparing the quoted Dhrystone MIPS—millions of instructions per second.

For advanced mathematical applications, required processing power will be more; hence, MPU is selected in such situation. If the application is real time in nature, then MCU will be the ultimate choice; MCUs with timing deterministic processor core and flash memory make them suitable for applications that need functional safety.

Memory

The next criterion for selection of MCUs and MPUs is based on memory availability on chip or external memory. To store and execute the program, MCUs usually have on-chip flash memory. This memory is embedded on the chip; it is difficult to add more memory if the code size exceeds. Flash memory's advantage is faster access time. If the on-chip memory is not sufficient, one can swap the device in same family with more memory.

For program and data storage, MPUs use external memory which offers lot of flexibility. External NAND or serial flash is often used to store the program, then it is loaded into external DRAM; hence, the start-up process takes longer time than MCUs which have embedded on-chip memory.

Power and Price

MCUs are clear winners over MPU as far as power consumption is concern. They have various modules available inside it, and if you are not using them in your application, those modules just go in idle mode and save lot of power. Designing application by keeping power consumption to the lowest value with an MPU is difficult and tricky. There are some MPUs which come with modes consuming low power, but these are few and are complicated to achieve.

A very important aspect in the performance–power trade-off is price. Obviously, the price of an MCU or MPU will have a big role to play in whether it is selected or not. Here, MCU is the more cost-optimized solution, and also the low-power option. But, does it have the performance required? An MPU is generally used for high-performance applications, but can you afford it? Designer must find answer for all these questions in order to make the choice.

Time to Market

To sustain in the competitive market, tight time-to-market deadlines with simplicity of design are very important. MCU needs only one power rail section, whereas an MPU core needs several different voltage rails, the DDR, and other blocks, so additional power converters are required, which further adds complexity and cost of the design.

Last but not least, sometimes it is required to modify the existing product, and planning for the future use is important. In these cases, selecting a vendor with an extensive range of MCU and MPU products that are compatible will help maximize software reuse when the time comes.

So, the solution to this is **programmable logic devices (PLDs)** which offer the flexibility of redesigning and upgrade the entire designed product without changing the platform.

1.2 Family Tree of PLDs

PLDs are categorized as: simple programmable logic devices (SPLDs) and high-density programmable logic devices (HDPLDs). SPLDs are further divided in the programmable array logic (PAL) and programable logic array (PLA) architecture, while high-density PLDs (HDPLDs) include complex programable logic device (CPLD) and field programable gate array (FPGA). Figure 1.1 gives the PLD tree diagram which is self explanatory.

Simple Programmable Logic Devices

Devices under SPLD are PALs and PLA. PLAs and PALs have packing density up to several hundred gates. The basic PALs architecture of AND/OR is implemented in sum-of-product form (SOP) using Boolean equations. PLDs' advantage is that in order to get higher packaging density, it replaces small- to medium-scale integrated (SSI/MSI) circuits. Single PLD device replaces IC with hundreds of equivalent gate. Another advantage of SPLD is that they consume very less power, fast performance; turn-around time is faster because of very few interconnects between the chips; and they are also highly reliable in nature. SPLDs are categorized under bipolar and CMOS technology. CPLD devices are higher in density, but SPLDs still have the best performance and easy to use.

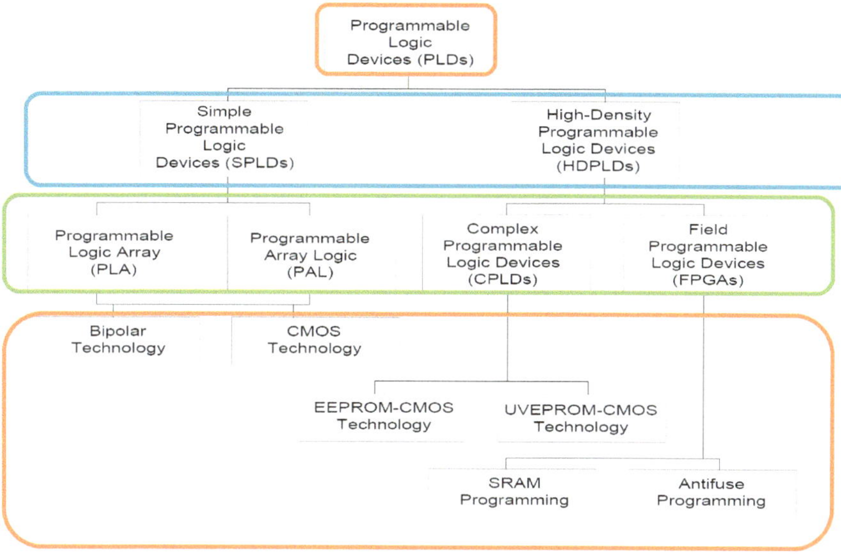

Fig. 1.1 Family tree diagram of PLD

New advancements in technology help SPLD to include programmable output logic, macrocells which can be configured in lower voltage and low power. This feature of SPLD allows more flexibility in design applications such as low power, high frequency, and low voltage which are most important.

High-Density Programmable Logic Devices

HDPLD has two high-performance devices namely CPLDs and FPGAs. The main drawback of SPLD is that it has limitation of architectural design such as only few logic structures that can be configured in a design and that to in a fixed defined way. HDPLDs on the other hand overcome the silicon scarcity by adding more blocks of flexible structures and interconnects. The two main components of CPLDs and FPGAs are the block interconnect and logic block elements. The other names for logic elements are logic cells/macrocells/logic blocks. The interconnects are nothing but how those logic elements are connected together to a desired design for a given application.

Both CPLDs and FPGAs are available in SRAM-based programming configuration, but only CPLD devices are EPROM or EEPROM programmed which means CPLDs' logic is not volatile. When FPGAs programming technology are antifuse-based which means one-time programmable (OTP) devices and SRAM-based means it can be programmed multiple times as and when required. This means that CPLDs can be up and running when power is applied and are nonvolatile.

Antifuse versus SRAM

The following is a list of advantages and disadvantages for the two technologies:

- SRAM programming technology is much slower than Antifuse programming technology that is due to interconnect RC.
- Silicon area per gate in Antifuse technology is more and routing becomes easier than in SRAM technology.
- Antifuse FPGA has disadvantage that they require many process layers and mask steps and requires high voltage programming transistors.
- Compared to Antifuse Technology ,SRAM-based technology provides higher capacity .
- SRAM based technology allows in-system programmability and the ability to reconfigure the design during the debugging stage while Antifuse technology is one-time programmable (OTP).
- The Major drawback of SRAM technology is its volatile nature meaning it has to be reprogrammed every time when power is turned off and on.

1.2.1 When to Choose a PLD?

1.2.1.1 Tips on Choosing PLA, PAL, CPLD, and FPGAs

There has always been a lot of confusion in programmers' mind over when to use a FPGA and when to use a CPLD. Here, we try to give solution based on application. For control circuits and state machine-based control logic, CPLDs are ideal choice. They have fast, predictable timing. It is very difficult to predict the data path delays in a FPGA. The greatest advantage of FPGAs is that it has fine logic blocks and a flexible architecture for implementation in data path designs, register-rich designs, control logic designs as well as arithmetic and logic functions.

CPLDs are very small devices with no dedicated internal RAMS or multipliers. FPGAs have internal RAMS and multipliers. CPLDs are very cheap, and FPGAs range from cheap to very expensive.

So, if you know it is a very simple design with no need for any serious maths or storage, stick with a CPLD. Otherwise, you will need an FPGA.

CPLD

Quick start
Non-volatile
Reconfigurable
Fixed Path Delays
Low/Medium size Packing .

FPGA

- Setup Time
- Volatile
- Reconfigurable
- Suited for large designs.
- High packing Density

1.3 Major Players in the Market and Their Product Specialties

There are multiple PLD players in the market, and their market share is dependant on the customer support provided and integrated development tool interface. To keep up the customer satisfaction is a big task for the PLD manufacturers; each manufacturer keeps on updating their products regularly to meet the market demand. Major players are Xilinx, Altera, Lattice, Quicklogic, etc.

1.3.1 Overview of Xilinx Products (www.Xilinx.com)

The world's largest provider of programmable FPGAs and SoCs that set industry standard for the lower cost, highest performance, and minimum power utilization is Xilinx.

Recent technological innovations have transformed Xilinx to integrate "All" forms of hardware, software, digital, and analog programmable technologies into its All Programmable FPGAs and SoCs.

Comparison between various series of Xilinx FPGA is shown in Table 1.1.

Xilinx has also come out with customized soft core processor such as PicoBlaze and Microblaze.

Table 1.1 Xilinx FPGA comparison

	Spartan6	Artix7	Kintex7	Virtex7	Kintex ultra scale	Virtex ultra scale
Logic cells	147,443	215,360	477,760	1,954,560	1,160,880	4,432,680
BlockRAM (Mb)	4.8	13	34	68	76	132.9
DSP slices	180	740	1,920	3,600	5,520	2,880
DSP (symmetric FIR)	140 GMACs	930 GMACs	2,845 GMACs	5,335 GMACs	8,180 GMACs	4,268 GMACs
Transceivers count (Gb/s)	8	16	32	96	64	120
Transceivers speed (Gb/s)	3.2	6.60	12.50	28.05	16.3	32.75
Transceiver total bandwidth (full duplex) (Gb/s)	50	211	800	2,784	2,086	5,886
Memory interface (DDR3)	800	1,066	1,866	1,866	2,400	2,400
PCI express	x1 Gen1	x4 Gen2	x8 Gen2	x8 Gen3	x8 Gen3	x8 Gen3
Analog mixed signal	–	XADC	XADC	XADC	System monitor	System monitor
AES configuration	Yes	Yes	Yes	Yes	Yes	Yes
Input/output pins	576	500	500	1,200	832	1,456
Input/output voltage (V)	1.2–3.3	1.2–3.3	1.2–3.3	1.2–3.3	1.0–3.3	1.0–3.3

Xilinx devices find use in number of broad range of applications such as surgery using robotic arm, Mars probes, wireless and wired network infrastructure, industrial automation, high-definition video, software-defined ratio platform, defense application.

1.3.2 Overview of Altera Products (www.altera.com)

Altera Corporation is the pioneer of programmable logic solutions, enabling system and semiconductor companies to rapidly and cost effectively innovate, differentiate, and win in their markets. Altera offers FPGAs, SoCs with embedded processor systems (NIOS II), CPLDs in combination with software tools, intellectual property, embedded processors, and customer support to provide high-value programmable solution.

FPGAs offer following design advantages in comparison to ASICs

- Quick Prototyping
- less time to market
- On the fly re-program capability
- Non Recurring Engineering costs is Low
- product life cycle is Long so risk of obsolescence is less

The Altera has developed FPGA Cyclone® series to meet programmers need of low-power, cost-effective design, reduce time to market. Every new generation of Cyclone FPGAs' series meets technical challenges of improved performance, increased integration, less power consumption, and quick time to market while meeting cost-effective requirements (Tables 1.2 and 1.3).

Table 1.2 Cyclone series

Series	Cyclone	Cyclone II	Cyclone III	Cyclone IV	Cyclone V
Year of introduction	2002	2004	2007	2009	2011
Process technology (nm)	130	90	65	60	28
Suited for new designs	Yes	Yes	Yes	Yes	Yes

Table 1.3 Arria family

Family	Arria GX	Arria II GX	Arria II GZ	Arria V GX, GT, SX	Arria V GZ	Arria 10 GX, GT, SX
Year of introduction	2007	2009	2010	2011	2012	2013
Process technology dimension (nm)	90	40	40	28	28	20

Lowest system cost and lowest power FPGA solution are provided by Cyclone V FPGAs for applications in the broadcast, consumer markets, industrial, and wireless sectors. This Cyclone family has integrated many hard intellectual property (IP) blocks which help to lower the system cost and also lower the design cycle time. The Cyclone V series offer customized SoC solutions in which Hard Processor System (HPS) ARM® Cortex™-A9 MPCore™ is present.

The market's low-cost, low-power FPGAs are Cyclone IV FPGAs, and now also they have a transceiver variant. Cyclone IV FPGA family is preferred due to its high-volume density and large bandwidth, keeping the system cost minimum.

Device family which offers power optimization, high functionality plus low power all in tandem is Cyclone III FPGAs.

Cyclone II FPGAs are designed to provide a customer-defined feature set and cost-sensitive applications. Its performance and low-power consumption is achieved at a cost lower than that of ASICs.

Initially, Altera brought Cyclone FPGAs series that was considered as low-cost FPGAs. But, today's designs require advanced features such as very low power and the higher packing density devices like Cyclone IV and Cyclone III FPGAs.

Altera's Arria® is midrange family which offers good power efficiency and optimal performance. ARM®-based hard processor system (HPS) is provided in Arria V and Arria 10 device families.

- Arria V GZ FPGAs support the maximum bandwidth compared to any 28 nm midrange FPGA.
- The Arria II FPGA family is based on a 40 nm, full-featured FPGA.
- Altera's cost-optimized 90 nm FPGA family with transceivers is Arria GX FPGA.

Products with high-performance, faster time to market, high productivity can be achieved by using Stratix® FPGA and SoC family. By using this highly rich feature set, Stratix FPGAs series (Table 1.4) help to increase system bandwidth and integrate many functions.

- Stratix 10 series FPGAs were introduced in 2013 with HyperFlex architecture encapsulated on the Intel 14 nm Tri-Gate process. It delivers double core performance and offers highest performance bandwidth and system integration.
- Stratix V and Stratix IV FPGAs provide the highest bandwidth, highest levels of system integration with 28 and 40 nm technology, respectively.
- Stratix III FPGAs are 65 nm introduced in 2006 for high-end sophisticated system processing designs for many applications.

Table 1.4 Stratix series

Device family	Stratix	Stratix GX	Stratix II	Stratix II GX	Stratix III	Stratix IV	Stratix V	Stratix 10
Year of introduction	2002	2003	2004	2005	2006	2008	2010	2013
Process technology dimension (nm)	130	130	90	90	65	40	28	14 Tri-Gate

- Stratix II and Stratix II GX variant has added an adaptive logic module (ALM) architecture, which helps in achieving high performance.
- Original members of the Altera® Stratix family are Stratix FPGAs and the Stratix with 130 nm technology.

1.3.3 Overview of Lattice (http://www.latticesemi.com/)

Lattice semiconductor brought their PLDs in market with low power, small form factor, low cost, customizable solutions for a quickly changing connected world. They are considered as leader in low-power design. Lattice semiconductor CPLDs use EECMOS technology which is non-volatile in nature. There are basically six families of Lattice PLDs such as 1000/1000E, 2000/2000V, 3000, and 6000. The packing densities of these families range from 1000 to 25,000 PLD gates. A very important feature of these PLDs is that it supports global routing pool, which helps to connect all internal structures and I/O's. Another important feature is the generic logic blocks (GLB).

Market for Lattice semiconductor FPGA is in the field of consumer appliance, communication, and industrial area.

1.3.4 Overview of QuickLogic (www.Quicklogic.com)

Initially, QuickLogic Corp. brought few FPGAs in the market, but due to lot of completion in market, they thought of backing away from the FPGA market, saying it will instead focus on an application-specific standard product (ASSP) and customer-specific standard products (CSSPs).

QuickLogic has been selling the PolarPro line of low-power, one-time programmable FPGAs, which competed against products from rivals Altera, Lattice, and Xilinx but could not sustain for quite long.

1.4 Overview of Software Tools

There are several books available in the market which explains in detail about an HDL languages. Here, we have attempted to give the glimpses of VHDL, Verilog, and ABEL hardware descriptive languages.

1.4.1 Programming Aspects of VHDL

What is VHDL?

To abbreviate VHDL, there are two parts V+HDL; V is nothing but VHSIC HDL. VHSIC is abbreviated as very high-speed integrated circuit. It helps to describe the functional behavior and structure of electronic design. The VHDL is regulated by the IEEE standards. VHDL language uses simulation and synthesis tools to design any system.

Design approach of VHDL is very flexible in nature; it does not constrain the user to specifically stick to one style of description. VHDL allows designer to describe the designs using top-down, bottom-up, or middle-out. VHDL even can be used to describe gate-level hardware. The most important feature of VHDL is that it helps to simulate the design before being sent for manufacturing, so that designers can quickly correct the designs and also do the functional simulation.

Design Flow using VHDL

Figure 1.2 shows the high-level design flow for FPGA. Following steps are needed to be followed for designing any system.

System-Level Verification

First step is system verification. Here, the entire system design having one or more devices is modeled and simulated using VHDL. Before commencing with detailed system design, its entire functional description of the system is validated.

Test bench Creation and RTL Design

The actual FPGA design commences once the entire architecture of the system is ready. This starts after putting the design at the register transfer level in VHDL, and

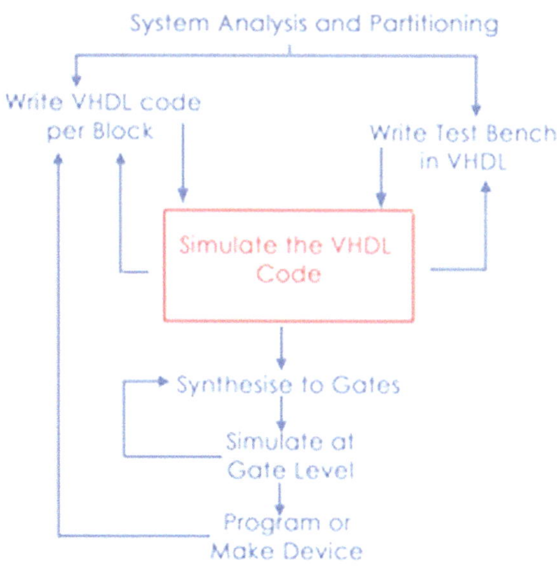

Fig. 1.2 Design flow of an VHDL

also capturing VHDL test cases. Both these tasks are exactly opposite and are sometimes performed by different design teams to ensure that the interpretation of specification is done correctly. The major task of engineers is to generate precise test cases to improve the quality of final FPGA/ASIC.

RTL Verification

System design functionality is validated against the specification by performing RTL VHDL simulation. RTL simulation is considered as faster than gate-level simulation.

Designer spends 70% of the entire design cycle in writing and simulating the design at register transfer level, and remaining 30% of the time is for verification and synthesis.

Look-ahead Synthesis

Before the actual synthesis, some exploratory synthesis will be done on the design process, which provides accurate speed and area. The main synthesis is not performed until the functional simulation is complete. It is not advisable to put lot of time and effort in synthesis before the functionality of the design is validated.

70% of design time at RTL! 30% of the time is for verification and synthesis. This is a rule to be followed. You must have heard of 20–80 rule in RISC, CISC design. Here, what we recommend is 70–30 rule.

A Simple VHDL Design Entity

A entire design entity is divided into two parts. In VHDL, they are called as design unit. External world interface to the design entity is defined by *entity declaration*. The *architecture body* defines the behavior and structure of the design entity, i.e., how inputs and outputs are related.

Here, we will describe a simple AND-OR-Invert (AOI) logic in VHDL as shown in Fig. 1.3. This design has four inputs and one output; we have taken into consideration the power and ground pins while modeling the design.

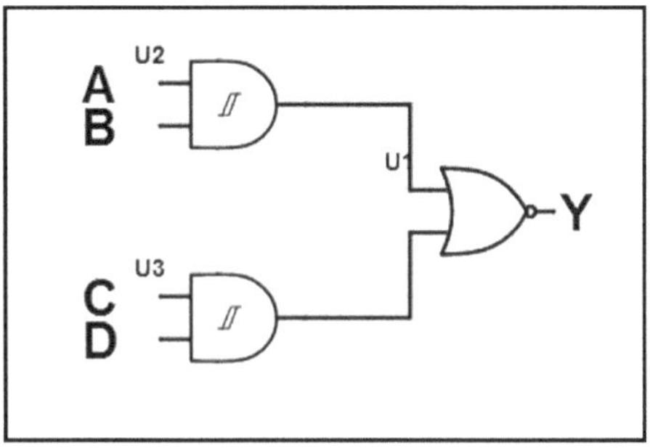

Fig. 1.3 Design entity of AOI gate

```
-- AND-OR-INVERT gate VHDL code

library IEEE;
use IEEE.STD_LOGIC_1164.all;

Entity AOInv is
port (
  A, B, C, D: in STD_LOGIC;
  Y : out STD_LOGIC
);
end AOInv;

Architecture V1 of AOInv is
begin
  Y <= not ((A and B) or (C and D));
end V1;
-- end of VHDL code
```

Detailed Line wise explanation of above code is given below.

Every programming language must have comment so that it is easier to understand the logic of code . Comment section starts with 2 hyphens mark (--) which is ignored by the compiler.

```
library IEEE;

use IEEE.STD_LOGIC_1164.all;
```

Library IEEE is always declared above the entity which helps to access the package STD_LOGIC_1164 of library IEEE for the declared name and data types in entity .

```
Entity AOInv is
```

Entity and is are the keywords of VHDL . Here AOInv is the name given for the entity by the user . The entity name is decided by the user.

```
port (

    A, B, C, D: in STD_LOGIC;
    Y : out STD_LOGIC
  );
```

Declaring entity means assigning name to the entity and port declarations. A port is basically input/output of the system. Each port declaration declares the direction the ports (in, out, or input–output).

The data type STD_LOGIC defines the set of values that may flow through the port, which is defined in STD_LOGIC_1164 in library IEEE package.

```
end AOInv;
```

keyword end is used to end the entity declaration.

```
architecture V1 of AOInv is
```

In the above line architecture, of and is are keywords of VHDL define in the packages. User can give any name to architecture body here it is given as V1. User can define multiple architecture bodies for a single entity. Name to the architecture is given to distinguish between multiple architecture declarations.

```
begin
```

begin is a keyword which tells this is the end of architecture declaration region and the start of statement portion of architecture.

```
Y <= not ((A and B) or (C and D));
```

The concurrent signal assignment in architecture describes the design entity function. The concurrent signal are executed when any one of the of the four ports A, B, C or port D changes value .

```
end V1;
```

The architecture is terminated by end keyword followed by the name of the architecture.

1.4.2 Programming Aspects of Verilog

Another hardware descriptive language is Verilog which resembles very close to C language. The IEEE standard 1364 is used to describe Verilog. There are basically three versions of Verilog; first version was published in 1995 and revision to this came in 2001 and 2005. Most of user uses Verilog 2005.

A Brief History of Verilog

Gateway Design Automation company way back in 1980s developed a logic simulator, Verilog-XL. Later in 1989, Cadence Design Systems acquired Gateway with its full rights to the language and the simulator. In 1990, Cadence put the language (but not the simulator) into the public domain, with the intention that it should become a standard, non-proprietary language.

Non-profit making organization, Accellera which was formed from the merger of Open Verilog International (OVI) and VHDL International maintains the Verilog HDL. OVI deals with IEEE standardization procedure.

Design Flow using Verilog

Figure 1.4 shows the high-level design flow for an FPGA using Verilog. The FPGA design flow steps using Verilog are same as that of design using VHDL, which also follows the 70–30 rule which is the rule of HDL.

Verilog-Based Simple Design

In Verilog, design is described by using the unit called as module. A module comprises of two parts, the declarations of port and the body of the module. The *port declarations* are same as entity declaration in VHDL which defines external interface of the module. The body of the *module defines the behavior and structure of the* design entity, i.e., how inputs and outputs are related. Let's us consider a simple AND-OR-Invert (AOI) logic in Verilog.

```
// AND-OR-INVERT Gate Verilog code
module AOInv (input A, B, C, D, output Y);
  assign Y = ~((A & B) | (C & D));
end module
// Verilog code end here
```

OK, that's the simple code for the design. Detailed Linewise explanation of above code is given below.

Comments

```
// Verilog code for AND-OR-INVERT gate
```

In Verilog, comments are entered by putting two forward slash marks (//). A comment line can be entered on a same line or on a separate line. Also, in Verilog, comment statements are ignored by the compiler.

Module and Port Declarations

```
module AOInv (input A, B, C, D, output Y);
```

Fig. 1.4 Design flow using Verilog

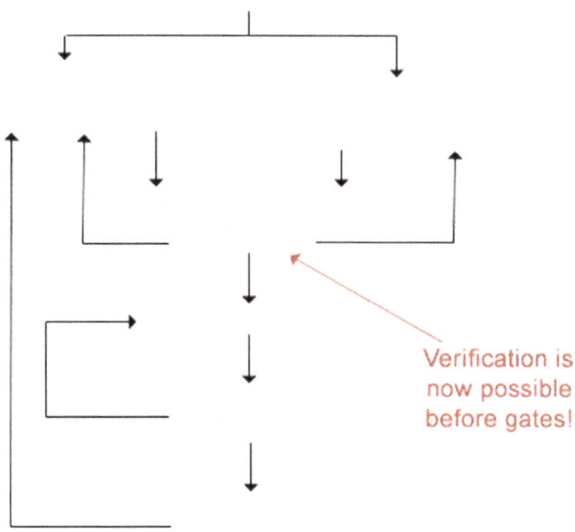

User can prescribe any arbitrary name to the module. Here, **module** is a Verilog keyword. New Verilog module definition is started with this line. Parentheses after the module name declare the input and output ports of the module. Port names (A, B, C, D) with their directions, i.e., (input, output, or input–output) are also declared.
End module
End module keyword terminates the module.

Functionality

Job of the module is to handle the interfaces, but how these ports are interrelated is defined by its functionality.
assign Y = ~((A & B) | (C & D));
Here, all the ports used are declared in the header port declarations section of module. Verilog keyword Assign is used to assign output of computed logic to the port declared as output. This also defines concurrent signal assignment, i.e., concurrent signal is executed when any one of the four ports A, B, C, or D changes value.

1.4.3 Programming Aspects of ABEL

The **Advanced Boolean Expression Language** (**ABEL**) is another hardware descriptive language. Data I/O Corporation has developed ABEL in 1983.

ABEL helps to describe any digital designs by equations, truth tables, state diagrams, or the combinations of all three. The main feature of ABEL is it helps to optimize and simulate the design without specifying a device or assigning pins. Test vectors description can also be given by ABEL.

With the advancement in field programmable gate arrays (FPGAs), standard hardware description languages (HDLs) such as VHDL and Verilog have gained in popularity PLD due to its large library support and resources. Also, ABEL still remains in use by thousands of PLD programmers worldwide.

Basic Structure of an ABEL-HDL File

ABEL-HDL follows the below syntax rules and restrictions:

- Maximum length of code line may be up to 150 characters.
- Code lines are terminated by either line feed character, form feed, or vertical tab.
- Keywords and numbers must be separated by at least one space.
- Identifiers are separated **by** commas.
- Keywords, numbers, operators, or identifiers cannot have spaces. Spaces are allowed in strings, explanation comments, and actual arguments.
- Case sensitive keywords and dentifiers in ABEL-HDL.

The syntax rules for identifiers are:

- Identifiers can be maximum up to 31 characters long.
- The first letter of identifier must start with an alphabet or with an underscore.
- Single names can have character “tilde” (~).
- Identifiers can contain upper, lower case characters, digits, and underscores in between except the first letter.
- Spaces cannot be used in an identifier.

The Source File Structure of ABEL-HDL:

Module is a unit which includes complete functional description of the design. An ABEL program supports multiple modules to be defined, but at a time, it will consider only the first module and the other modules will be checked according to the rules of syntax. One module can only have one PLD device specification.

ABEL programming module consists of the following flow setup:

Header
Declarations
Logic Description
Test Vectors
End

The module structure follows below four rules:

- Only one header should be present in a module.
- Declarations must immediately follow the header.
- Other sections of a source file can come in any order.
- Identifier or symbols cannot be referenced before it is being declared.

Chapter 2
Getting Hands on Altera® Quartus® II Software

Contents

Abstract This chapter provides users with overview and capabilities of Altera® Quartus® II design software in programmable logic design. The book is designed around the Altera DE2 development platform. The Altera Quartus II software is the most comprehensive environment available for system-on-a-programmable-chip (SOPC) design. Here we provide a guide that will help one to install the Quartus software, setting up the license for installed Quartus. This chapter also gives the details about the steps involved in creating the first embedded project, building projects' steps, and how to port the programming file onto the development board.

Keywords Altera · Quartus II · SOPC · Embedded platform

This chapter provides users with overview and capabilities of Altera® Quartus® II development software tool in programmable logic design. Quartus II software platform is more suitable for system-on-a-programmable-chip (SOPC) design in many applications. This chapter will give simple and easy steps to user about how to install software setup and the license creating first embedded design and simulation of entire design.

J.S. Parab et al., *Hands-on Experience with Altera FPGA Development Boards*,
https://doi.org/10.1007/978-81-322-3769-3_2

2.1 Installation of Software

This chapter gives the minimum requirements of the system and installing procedures of Quartus II software.

System Requirements

Following minimum system requirements need to be verified before Quartus II software installation.

- Pentium II PC runs at more than 400 MHz with at least 256 MB RAM
- Disk space 1 GB where you are installing the Quartus II software
- Windows NT version 4.0, Windows 2000, or Windows XP
- CD-ROM drive
- Following ports availability:
  - Parallel port for ByteBlaster™ II or Byte BlasterMV™
  - Serial port for Master Blaster™
  - USB port for USB-Blaster™
- Internet Browser, i.e., Internet Explorer 5.0 or later

Uninstalling Previous Versions of Quartus II Software

Before starting with the fresh installation of Quartus II, it is recommended by Altera to uninstall the previously installed version of Quartus II by following below steps. Sometimes user can install new version without removing the old one by simply specifying the new installation directory.

Choose Start > All Programs > Altera > Quartus Uninstall, Repair or Modify.

Running the Setup of Quartus II

Following steps need to be followed to install the Quartus on user system:

Quartus software installation is allowed only if user has administrator privileges.

1. Insert CD having Quartus II Software into CD-ROM drive. Immediately, several installation options pop up and one can click on install option. Setup can also be started manually by performing the following steps:
   a. Start > Run.
   b. In the dialog box, type <CD-ROM drive>:\install.
   c. Press OK.
2. After clicking on Install Quartus II Software, installation starts automatically and guides user during installation process.

2.2 Setting Up of License

Licence.dat file needs to be obtained before setting up of license for Quartus II.

Steps to Obtain the license file

(1) Browse the portal address www.altera.com/licensing.
(2) Select **Get licenses** link which is the first blue link on the page.
(3) Press on **Get a license for Quartus® II Web Edition software**.
(4) One can get one time access if you create an account using you email id.
(5) Once you create username, go back to Step 2.
(6) To get license you have to provide system network interface card number (NIC).

NIC number is a twelve digit hexadecimal number that recognizes your system, and one can easily find the NIC of system by typing `ipconfig/all` at a windows command prompt.

To obtain NIC number, following command typed on command prompt window:

`Ipconfig/all`
Search for following line
`Physical Address .................00-ED-6C-59-91-4F`

Then on licensing tab, user has to enter the above NIC without the (-).

Once you submit the required information, the license file will be emailed to you.

Once the user receives license file, next step is to do the license setup if the user has Windows XP system, and then specify the location of the license file by following the below steps:

i. Choose start > Control Panel.
ii. Click on System in Control Panel window.
iii. Then click on Advanced tab > click Environment Variables.
iv. Click on New tab Under System Variables. Then new system variable dialog box appears.
v. Specify Variable Name as LM_LICENSE_FILE.
vi. In the value box, type either <drive>:\flexlm\license.dat or server host name <port>@<host>. If there are more than one license file or server, separate the port and host specifications with semicolons (;), with no spaces between the names and numbers.
vii. Click OK.

2.3 Creation of First Embedded System Project

Here we will briefly introduce the Quartus II CAD for simple system. CAD flow is used for designing circuits which are implemented in FPGA using Quartus II software. How to use Quartus II Software for creation of simple embedded project on FPGA is explained in this section. The tutorial explained here uses VHDL to create design entry; here user clearly defines the desired circuit in the VHDL. Two other methods are Verilog description and Block Design file (BDF).

Background

A FPGA CAD flow is illustrated in Fig. 2.1. CAD tools make it easier for designer to implant the desired logic.

Fig. 2.1 FPGA design CAD flow

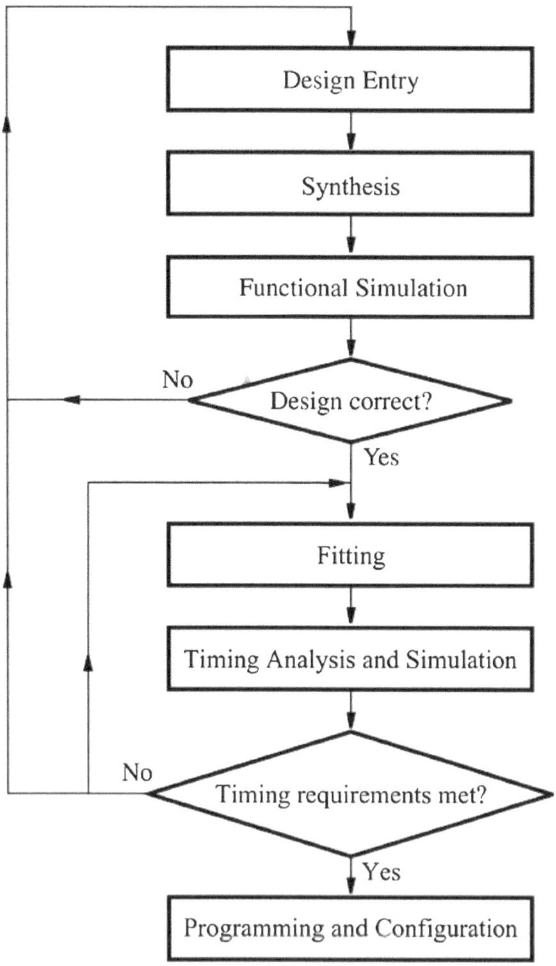

Following are the steps of CAD design flow:

- Design entry—Here user defines the circuit in BDF or HDL, i.e., VHDL or Verilog.
- Synthesis—The entire design is synthesized into a circuit which consists only logical elements.
- Functional simulation—This verifies the functional correctness; Here timing issues are not taken into consideration.
- Fitter—This tool helps to find the exact placement of LEs in FPGA as specified in the Netlist; it also selects wire routings to make the necessary connections of specific LEs.
- Timing analysis and simulation—Propagation delays along the various paths in the fitted circuit which are analyzed to provide an indication of the expected performance of the circuit.
- Timing—The fitted circuit is tested to verify both its functional correctness and timing configuration.
- Programming—The designed design is downloaded on the FPGA chip by programming which configures the various logical elements (LEs).

Embedded project design for three-input AND Gate

Here we consider simple three-input AND Gate, and detail steps to create a project are given below.

Logic Block Diagram

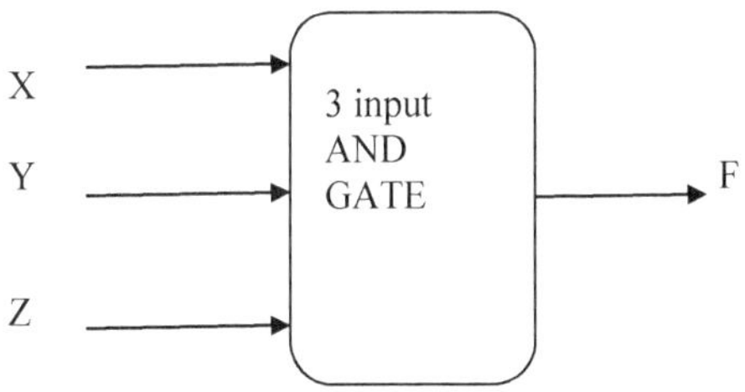

TRUTH TABLE

```
 +---------+--------+
 ¦ INPUT  ¦OUTPUT ¦
+---------+--------+
--¦    X Y Z ¦    F  ¦
-- +---------+--------+
-- ¦  0 0 0  ¦   0   ¦
-- ¦  0 0 1  ¦   0   ¦
-- ¦  0 1 0  ¦   0   ¦
-- ¦  0 1 1  ¦   0   ¦
-- ¦  1 0 0  ¦   0   ¦
-- ¦  1 0 1  ¦   0   ¦
-- ¦  1 1 0  ¦   0   ¦
-- ¦  1 1 1  ¦   1   ¦
-- +---------+--------+
```

VHDL code

```
library IEEE;
use IEEE.STD_LOGIC_1164.ALL;
use IEEE.STD_LOGIC_ARITH.ALL;

entity and3 is
port (X,Y,Z: in STD_LOGIC;
     F: out STD_LOGIC);
end  and3;

architecture BEHAVIORAL of and3 is
begin
 F <= (X AND Y AND Z);
end BEHAVIORAL;
```

(1) Goto Start > All programs > Altera > Quartus II 7.2 sp3 web edition the following window appears.

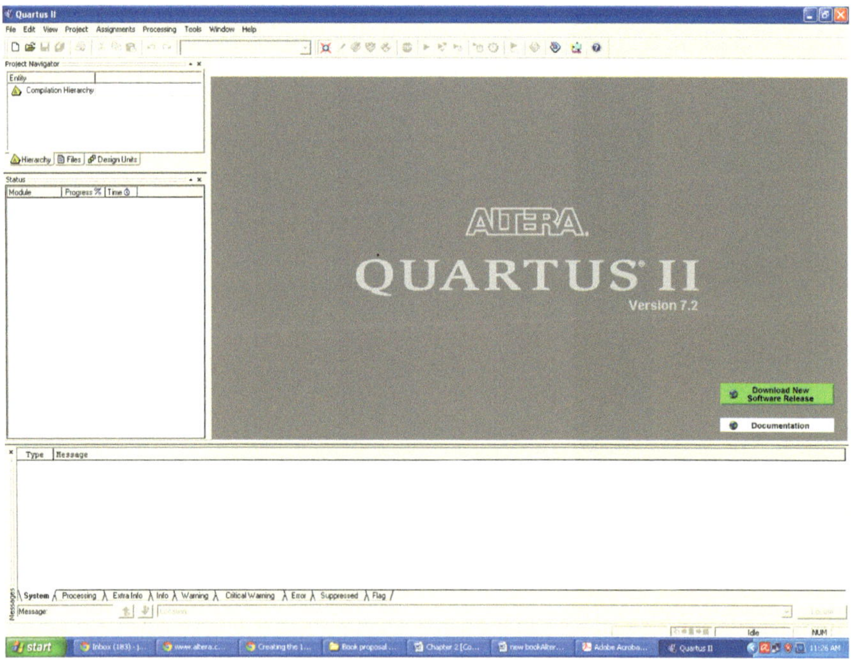

(2) Once the licensing setup is done, you can start working on Quartus by creating a project. Click on File → Create New Project wizard.
(3) The window shown in Fig. 2.2 pops up and one has to enter the directory to save project files, then assign name to file and a project, and click Next. (Note: top-level entity name must be same as file name).
(4) Then click Next in the following popup that comes, and it says directory does not exist. Do you want to create it? Say yes.

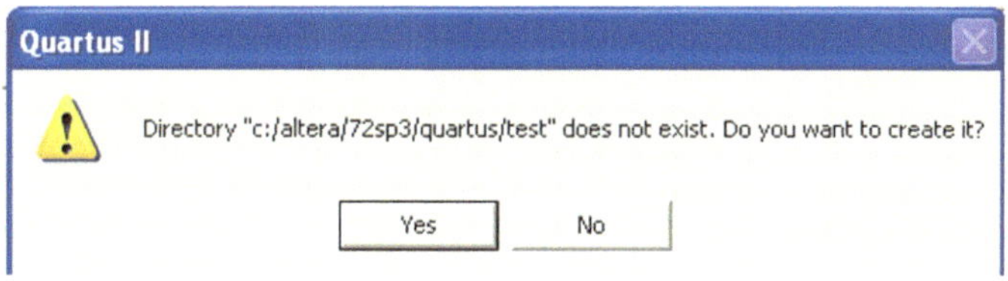

(5) Next window appears which will ask you to provide filename, do not assign any file name at this point just click on next Fig. 2.3 appears which will ask you to select the FPGA Family and Target on your respective Development Board (Cyclone II EP2C35F672C6).
(6) Click Next two times, you will get summary of project (Fig. 2.4) which complete the project creation. Then click finish.
(7) Select File > click on New, the window shown in Fig. 2.5 appears, select VHDL File, and press OK.

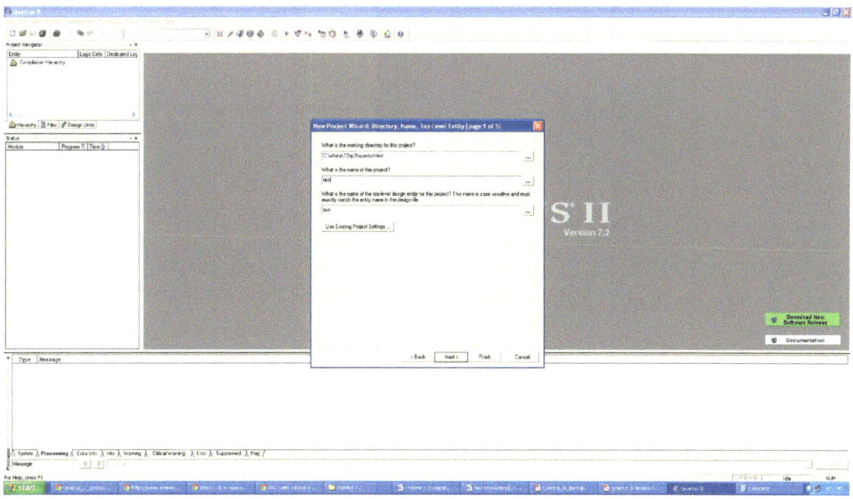

Fig. 2.2 Top level entity design name

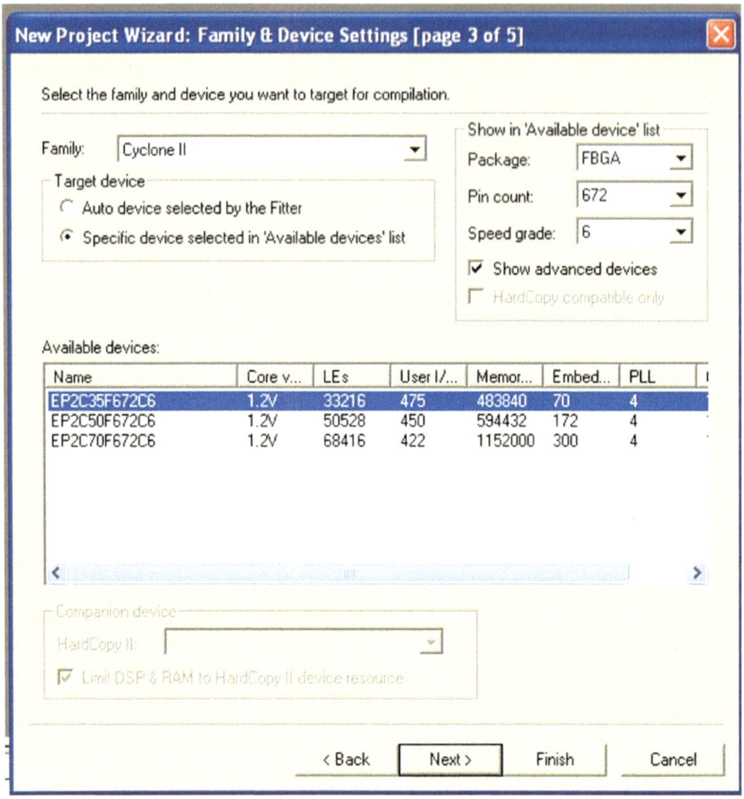

Fig. 2.3 Select FPGA device

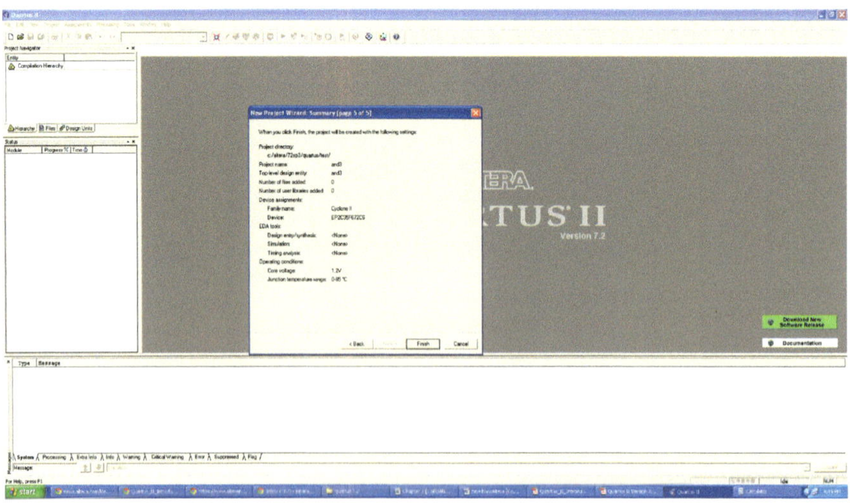

Fig. 2.4 Summary of project settings

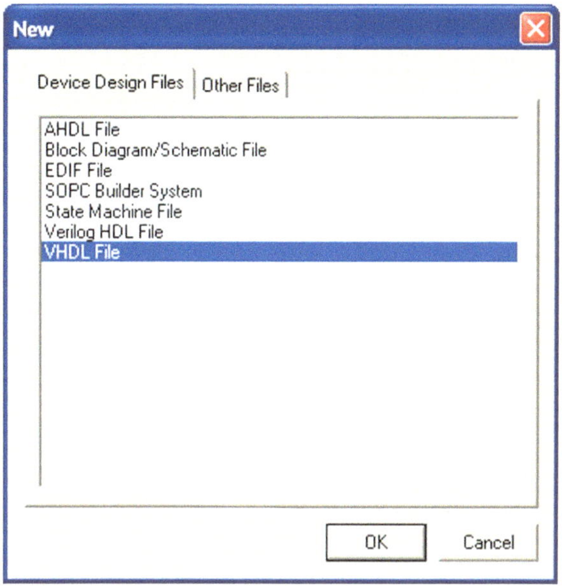

Fig. 2.5 Choosing VHDL file

(8) Then type in your logic code in the editor (Fig. 2.6) then go to file > save as, assign the file name same as entity name.

(9) **Next step is assignment of FPGA pins.**

There are two ways of assigning the pins, manual pin assignment and automatic pin assignment:

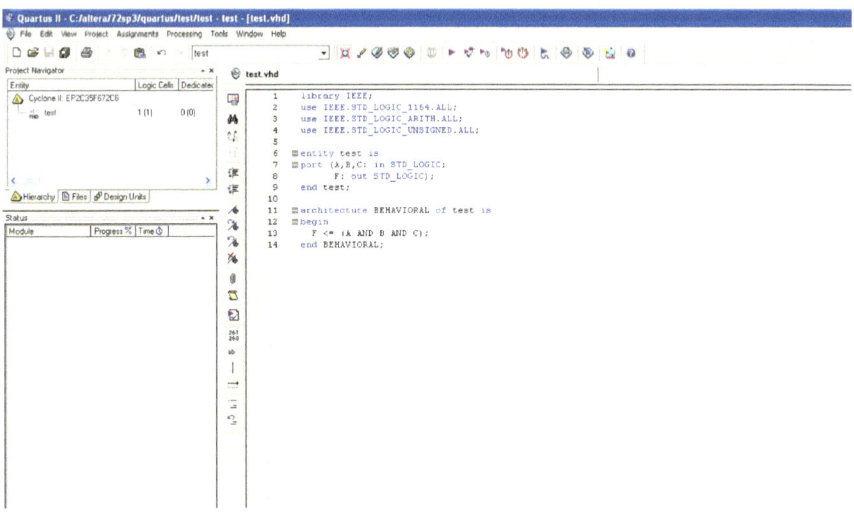

Fig. 2.6 Three-input adder VHDL code

(1) Manual Pin Assignment:

Here to see the pins in Assignment editor, directly one has to compile the entire system by clicking the start compilation under the processing toolbar; once the entire system is compiled without any errors (warnings generated are accepted), then go to Assignment → Pins.

Assign the respective pins of input to switches and output pins to LEDs.

(2) Automatic Pin Assignment

- Store the DE2 board pin assignment excel file on the computer
- Click on Project → Import design partition then select the location of pin assignment file stored on computer → click OK
- Go to project → generate tcl script for project then click OK.

2.4 Project Building and Compilation

(1) After completing the design, next step is to compile the design for errors.
(2) Click on Processing tab → start compilation This is shown in Fig. 2.7.

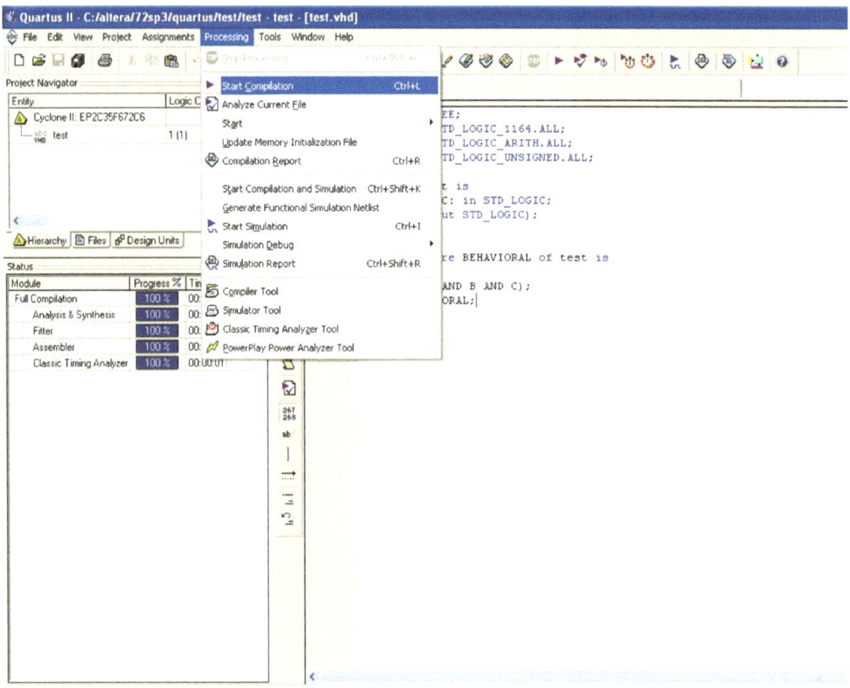

Fig. 2.7 Compilation of project

(3) After successful analysis and synthesis, compilation report is generated as shown in Fig. 2.8. If there are errors, then click on that error so that it helps you for debugging the design.

(4) If you want to see the designed system at Netlist (Gate) level, on the menu bar select tools > click on Netlist Viewer > RTL Viewer as shown in Fig. 2.9.

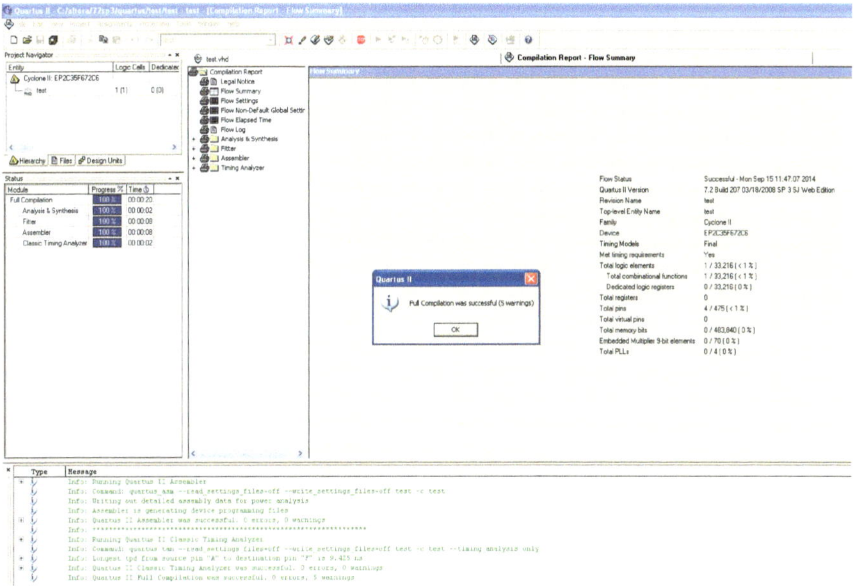

Fig. 2.8 Report of compilation

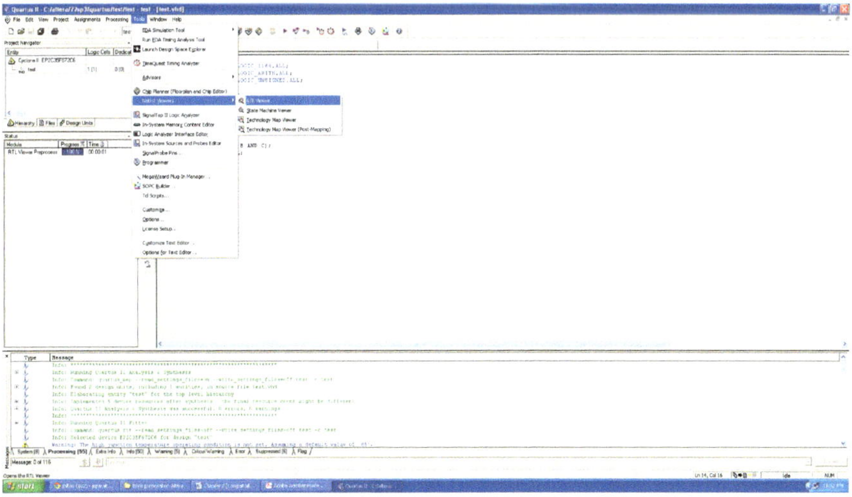

Fig. 2.9 RTL viewer selecting window

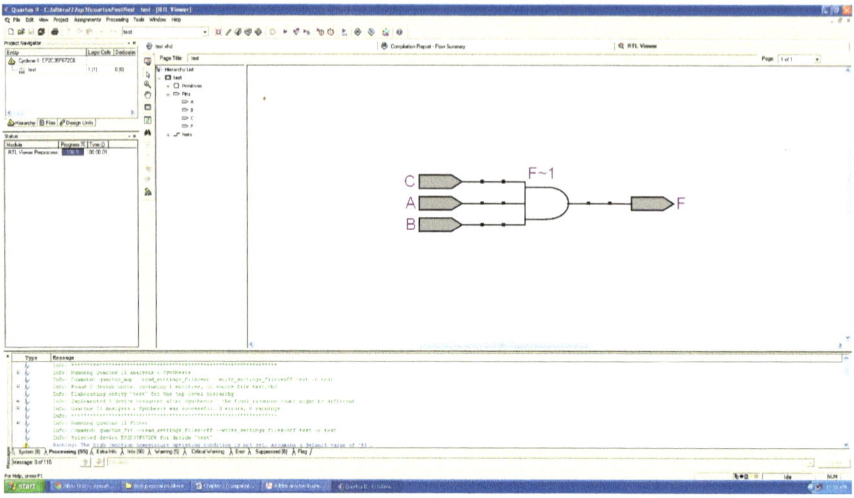

Fig. 2.10 RTL schematic of three-input AND Gate

After clicking on RTL, viewer, it will show Gate-level architecture as shown in Fig. 2.10.

Simulation of designed system

After successful compilation, the designed system can go for simulating the waveforms.

Select **File** → then click on **New** → **Other Files** → **Vector Waveform Files**, the window shown in Fig. 2.11 appears.

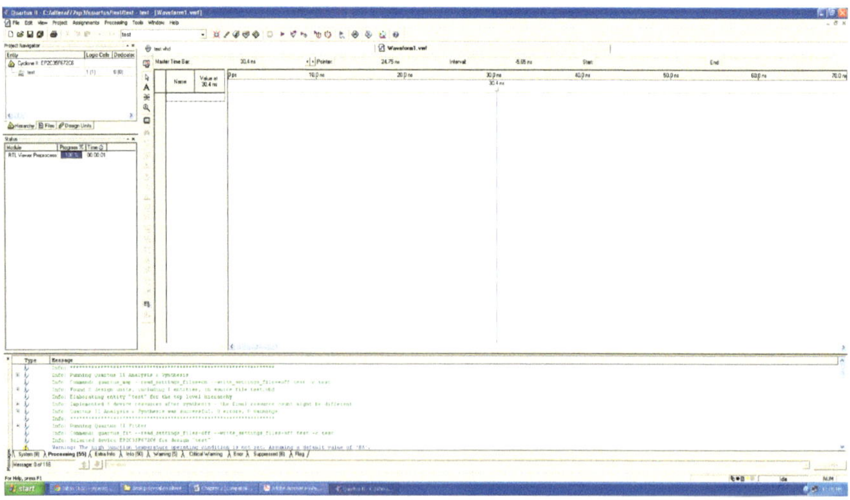

Fig. 2.11 Window for editing waveform

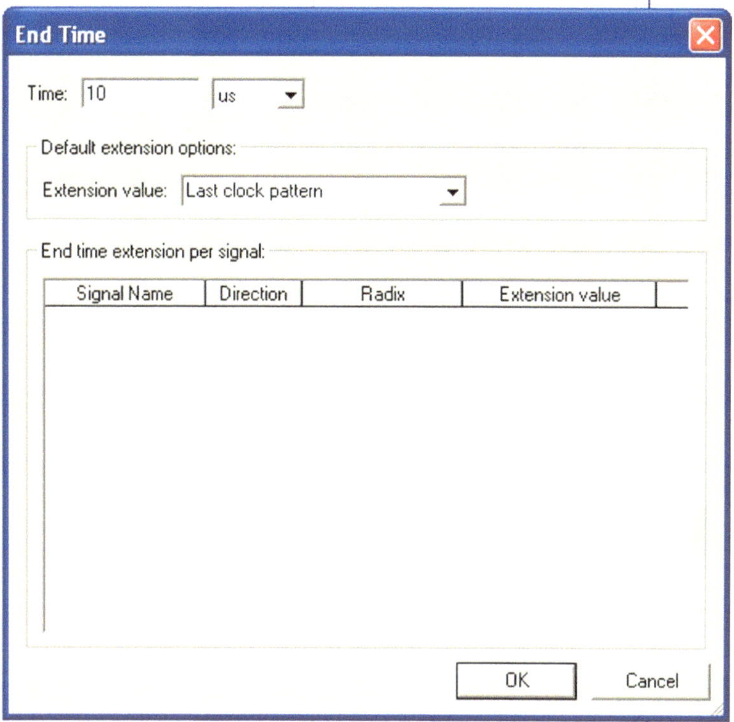

Fig. 2.12 Specifying ending time

Save the waveform with .vwf extension in the same project directory.

One can specify ending time for the waveform to be simulated by clicking on **Edit → End time** as shown in Fig. 2.12.

- To see the full simulated output waveform, click on **View → Fit in Window.**
- **To specify the input/ output nodes, click on Edit → Insert Node or Bus**. Figure 2.13 shows window of node finder.
- Select pins: all under filter and click on list and click on list.
- Clicking on list will give all the used pins in left pane. Select all the pins and click on > to move the pin to the right pane.
- Now, you get all the pins in your waveform editor window shown in Fig. 2.14.
- Click on overwrite clock to specify time period for input node from the vertical tool bar as shown in Fig. 2.15.

Let us specify time periods for inputs X, Y, Z as 1, 0.5 and 0.25 microseconds respectively as shown in Fig. 2.16. After specifying the time periods simulated inputs signal looks like as shown in Fig. 2.17.

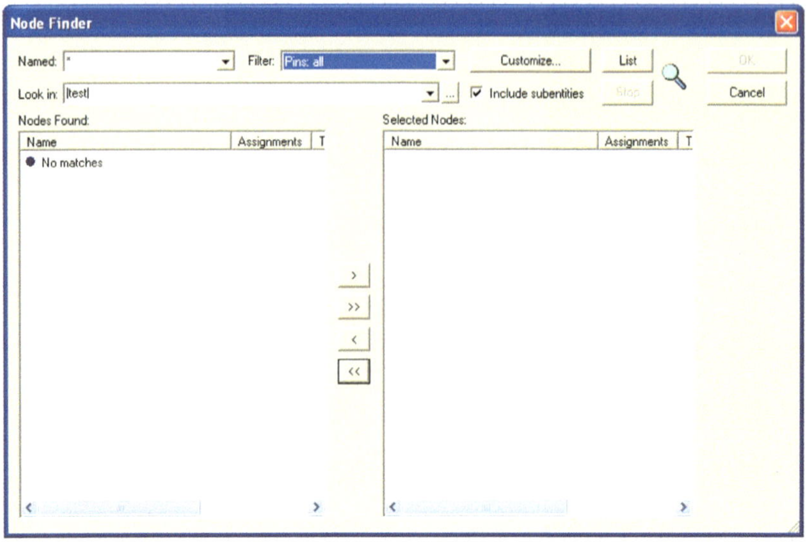

Fig. 2.13 Node finder window

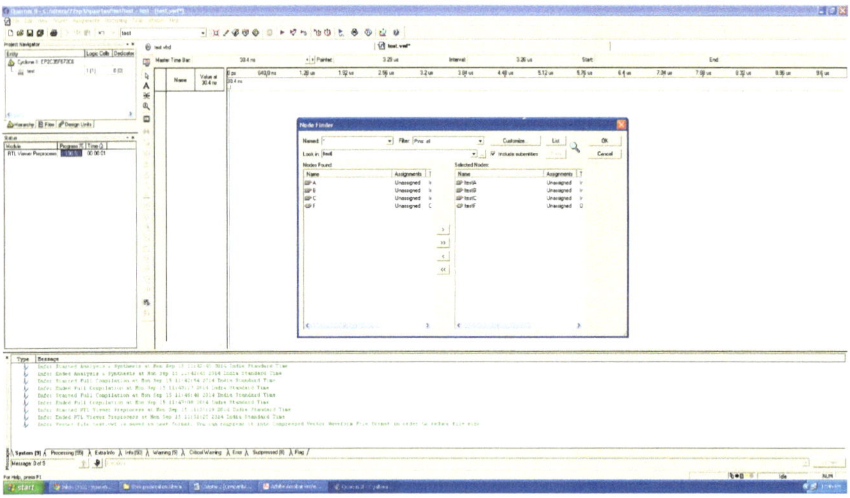

Fig. 2.14 Pin list window

- Quartus II simulator tool supports functional and timing simulation. Simulation mode is selected by clicking **Assignment** → **Settings** → **Simulator**. The screen should look like Fig. 2.18.
- Click on **Processing tab** → **click on Generate Functional Simulation Netlist**.

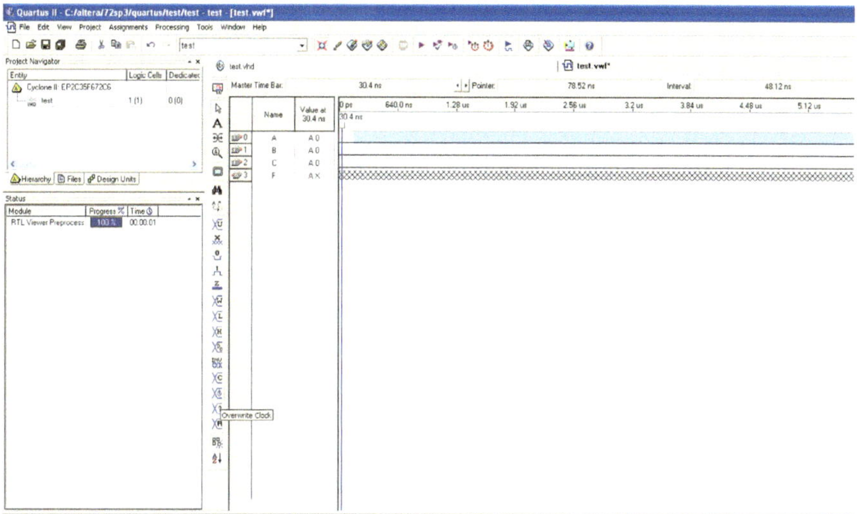

Fig. 2.15 Time period specifying using overwrite clock

Fig. 2.16 Period and start time entry

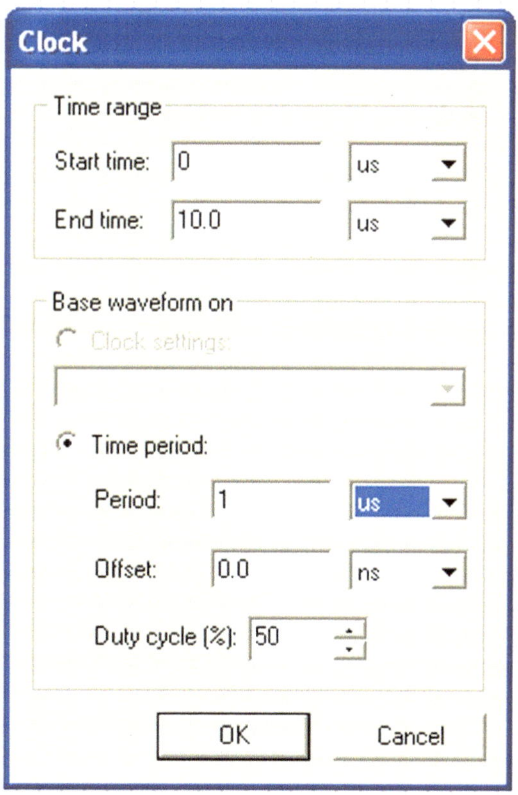

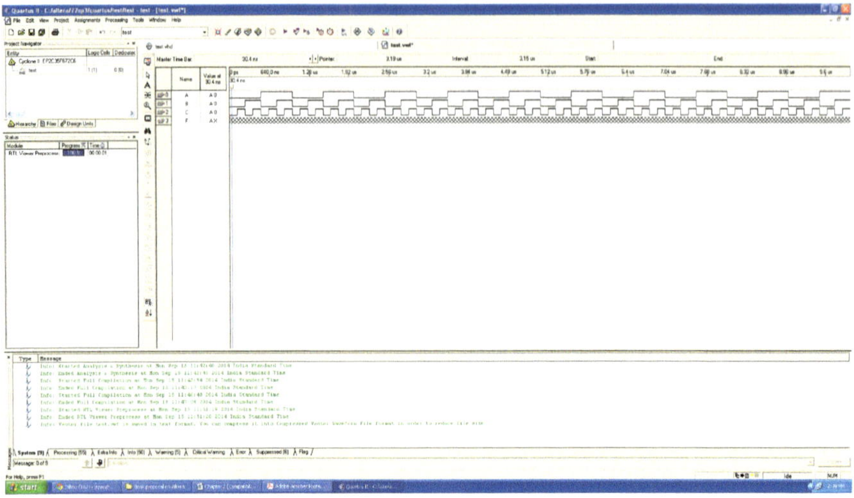

Fig. 2.17 Simulated input waveforms

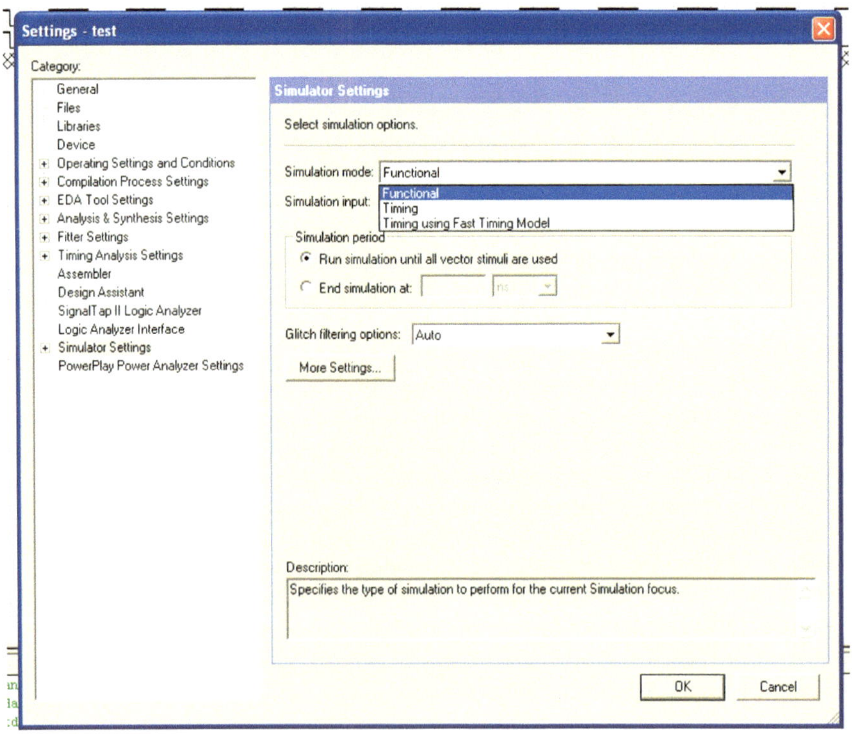

Fig. 2.18 Settings of functional simulation

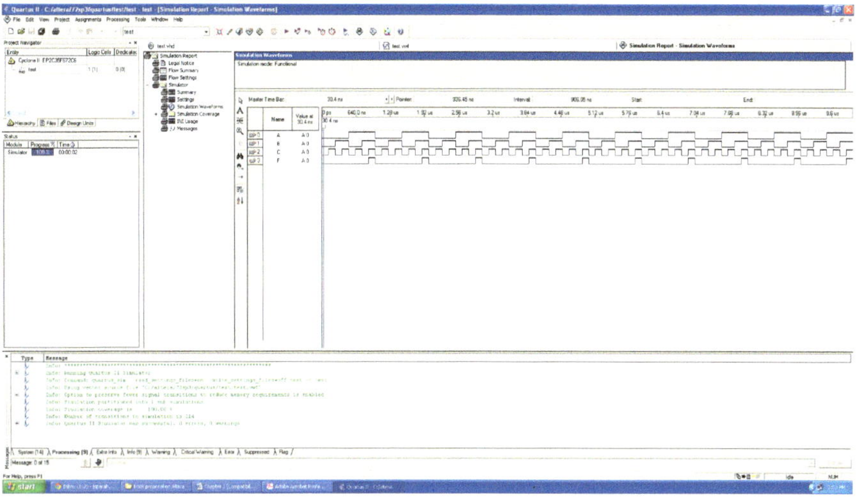

Fig. 2.19 Simulated waveform

- Then click on **Processing tab → select Start Simulation.** The final simulated output of three-input AND Gate is shown in Fig. 2.19.

2.5 Programming and Configuring the FPGA Device

To implement the designed system on FPGA, it is necessary to program the FPGA. Altera's DE-series board supports two different configurations, i.e., JTAG and AS modes. The file containing configuration data is downloaded from host PC to the board by using USB port. To use port connection, one has to install USB-Blaster driver.

FPGA devices are programmed directly by using JTAG mode. If the FPGA is programmed using JTAG mode, then it will retain its configuration as long as the power remains turned on. The configuration data is automatically erased when there is no power. The second configuration mode is active serial (AS); here the device is provided with some memory to load the configuration data file which is later on loaded on the FPGA device upon power-up. The selection between the two modes is made done by RUN/PROG switch on the DE-series board. JTAG configuration mode is selected by RUN position switch, while AS mode is selected by putting switch in PROG position.

Steps for programming and hardware setup

Click on Tools > programmer, the following window shown in Fig. 2.20 appears and one should do the hardware setup before downloading the configuration design file.

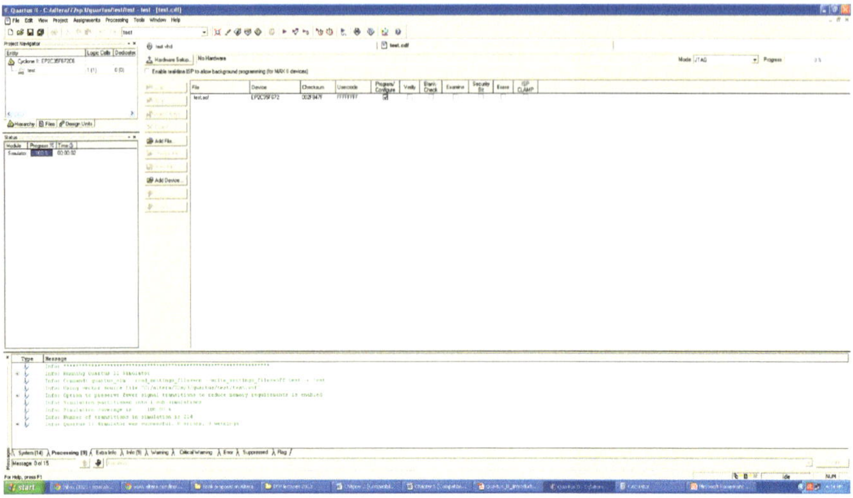

Fig. 2.20 Programming and configuring FPGA

USB-Blaster Driver Installation

One cannot proceed with programming with FPGA unless USB-blaster drivers are installed. Follow below steps for USB-Blaster driver on platform having Windows 2000 and Windows XP.

- USB-Blaster driver is present where Quartus is installed, i.e., at C:\altera\72sp3\Quartus\drivers\usb-blaster.
- Follow the below steps to install the USB-blaster driver:

  1. Connect the USB-Blaster download cable to the USB port of PC.
  2. You will see the Found New Hardware Messages gets pops up, click No.
  3. Select the option, Install from a specific location list and then click Next.
  4. Click on choose the driver and then click Next.
  5. Choose Altera USB-Blaster > click Next to continue.
  6. Click Next to install the driver.
  7. Press Continue Anyway if the warning message appears.
  8. Click Finish in the completing hardware installation.

USB-Blaster Hardware setup

To setup the USB-Blaster hardware, follow the below listed steps:

1. Open Quartus II software.
2. Click on Programmer (Tools menu).
3. After clicking on Hardware Setup the below window appears.

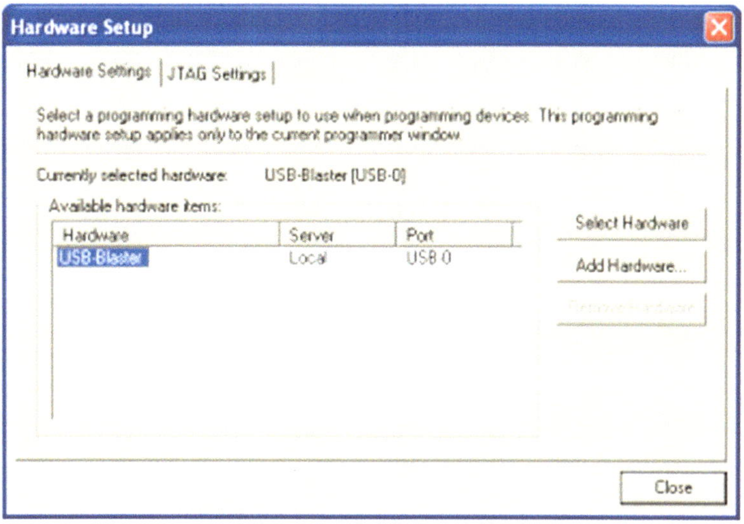

4. Select USB-Blaster from the drop-down menu.
5. Then Click on Close.

Following above five steps will complete the USB-Blaster setup.

Next step is to tick mark (√) the Program/Configure option as shown below and click on start, and this will start the device programming and shows the status of programming on progress bar.

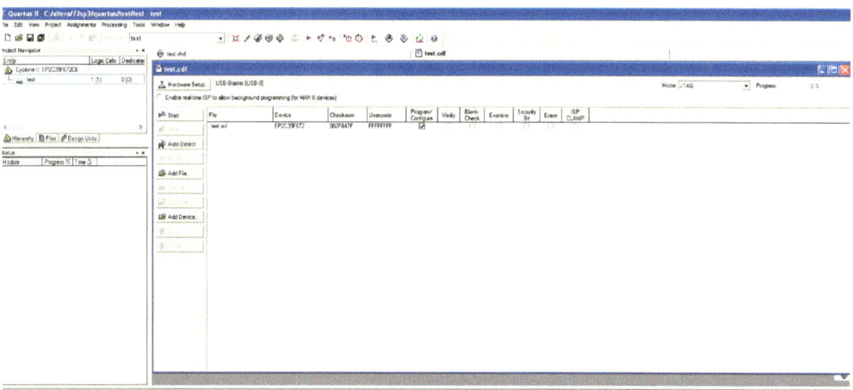

Once the programming is done, then next part is to test the logic onto the DE2 Board.

Chapter 3
Building Simple Applications with FPGA

Contents

Abstract This chapter gives the detailed implementation steps of six digital logic design applications such as multiplexer/demultiplexer, encoder/decoder, shift register, counter, memory, and traffic light controller. Here, the detailed steps are shown right from the creation of project till the simulation and final result on the development board.

Keywords Digital logic design · Vector waveform · RTL viewer

3.1 Implementation of 8:1 Multiplexer

Introduction

A combinational circuit which is used for transmitting large numbers of information to a small number of channels or lines is called multiplexer. This circuit is mostly used in application based on digital design.

A digital multiplexer is a device which selects one of the input lines out of many inputs and connects the selected input line to single output line. Input lines are selected by using select line. The block diagram and logic diagram of 8:1 multiplexer are shown in Figs. 3.1 and 3.2, respectively.

J.S. Parab et al., *Hands-on Experience with Altera FPGA Development Boards*,
https://doi.org/10.1007/978-81-322-3769-3_3

Design Description

A n set of select inputs are required to select any one input out of "m" inputs to connect to the output.

$$2^n = m$$

(m = number of inputs and n = number of select lines).

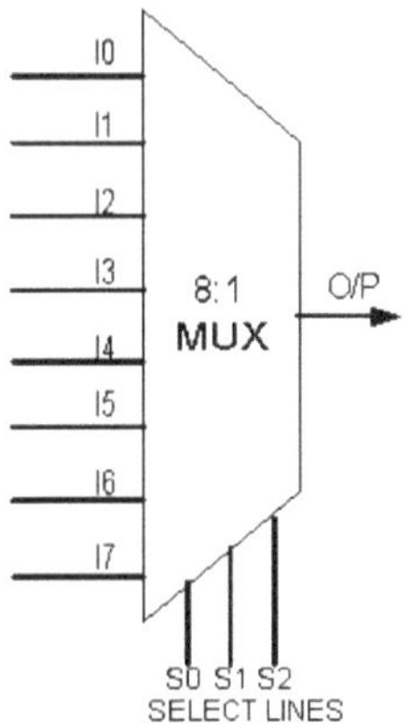

Fig. 3.1 8:1 multiplexer block diagram

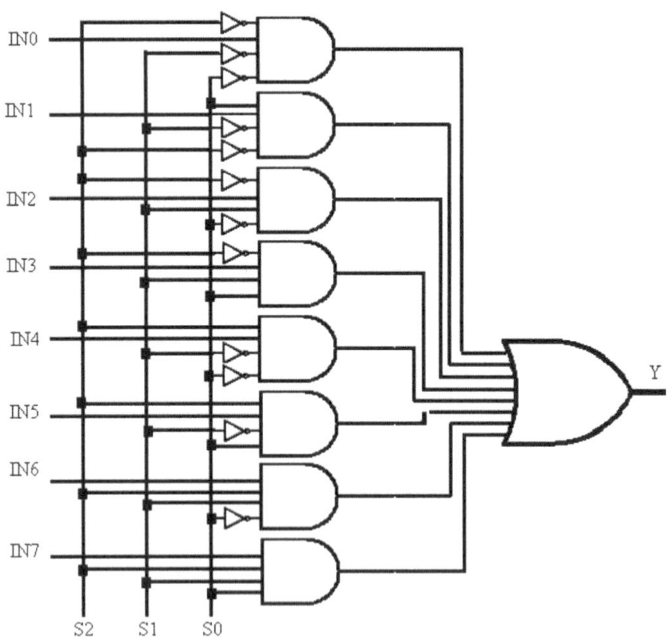

Fig. 3.2 8:1 MUX logic diagram

8:1 Multiplexer Truth Table

Inputs to be selected			Output
A	*B*	*C*	*Y*
0	0	0	*I*0
0	0	1	*I*1
0	1	0	*I*2
0	1	1	*I*3
1	0	0	*I*4
1	0	1	*I*5
1	1	0	*I*6
1	1	1	*I*7

Design Implementation Procedure

Step 1 Double click on Quartus—II 7.2.

Step 2 Go to file → New Project wizard and then click Next, and then, one has to give the working directory name (e.g., mux), project name, and top-level entity name as shown in Fig. 3.3.

Step 3 Then click Next 2 times which will ask to specify the device (Cyclone II EP2C35F672C6) as shown in Fig. 3.4.

Step 4 Type your code for 8:1 multiplexer and save file with the same name as your entity.

```
----------------------------------------------------------------------------
-- Title      : 8:1 Multiplexer
-- Design      : vhdl_test
-- Author      : Dr.J.S.Parab
----------------------------------------------------------------------------

LIBRARY IEEE;
USE IEEE.STD_LOGIC_1164.ALL;
USE IEEE.STD_LOGIC_ARITH.ALL;
USE IEEE.STD_LOGIC_UNSIGNED.ALL;

ENTITY MUXP IS
PORT ( SEL: IN STD_LOGIC_VECTOR(2 DOWNTO 0);
A,B,C,D,E,F,G,H :IN STD_LOGIC;
MUXP_OUT: OUT STD_LOGIC );
END MUXP;

 ARCHITECTURE BEHAVIORAL OF MUXP IS

BEGIN
```

```
PROCESS (SEL,A,B,C,D,E,F,G,H)

BEGIN

CASE  SEL IS

WHEN "000" => MUXP_OUT <= A;
WHEN "001" => MUXP_OUT <= B;
WHEN "010" => MUXP_OUT <= C;
WHEN "011" => MUXP_OUT <= D;
WHEN "100" => MUXP_OUT <= E;
WHEN "101" => MUXP_OUT <= F;
WHEN "110" => MUXP_OUT <= G;
WHEN "111" => MUXP_OUT <= H;
WHEN OTHERS => NULL;
END CASE;
END PROCESS;

END BEHAVIORAL;
```

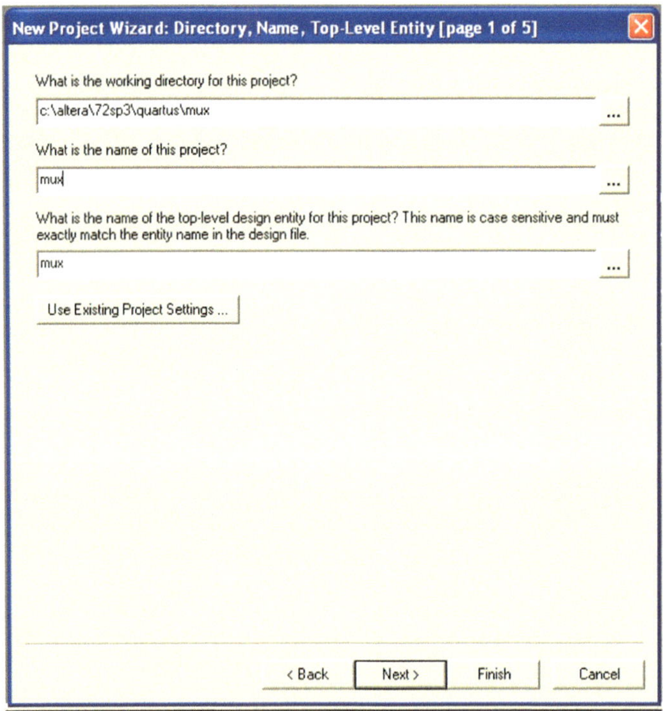

Fig. 3.3 Creating new project

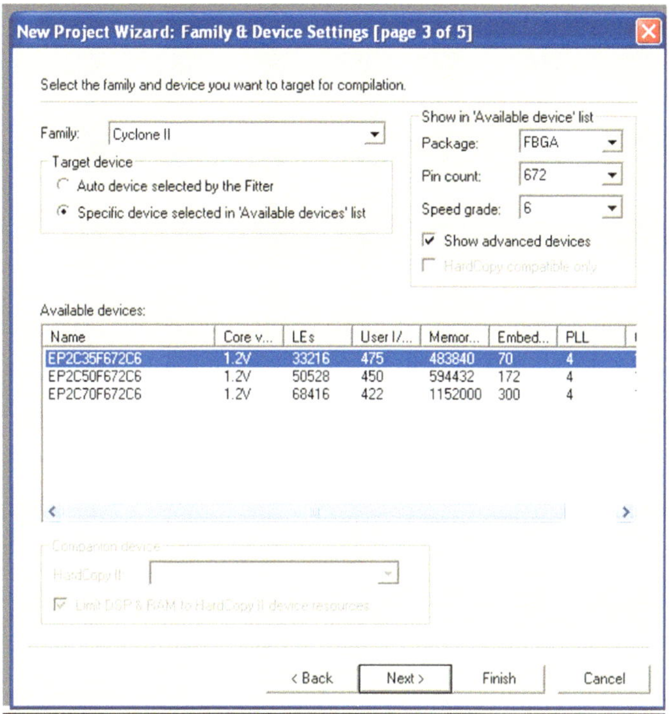

Fig. 3.4 Specifying the device target

Next Step is Assignment of FPGA Pins

There are 2 ways of assigning the pins: manual pin assignment and automatic pin assignment:

Manual Pin Assignment:

Here to see the pins in assignment editor directly, one has to compile the entire system by clicking the start compilation under the processing toolbar. Once the entire system is compiled without any errors (warnings generated are accepted), then go to Assignment → Pins.

Assign the respective pins of input to switches and output pins to LEDs.

Building and Compilation of Project

1. After completing the design, one can check the design for errors by compiling.
2. Compilation is started by clicking on Processing → start compilation.

After successful analysis and synthesis, compilation report is generated as shown in Fig. 3.5. If there are errors, click on that particular error to get more information or press F1. That will help you with debugging process.

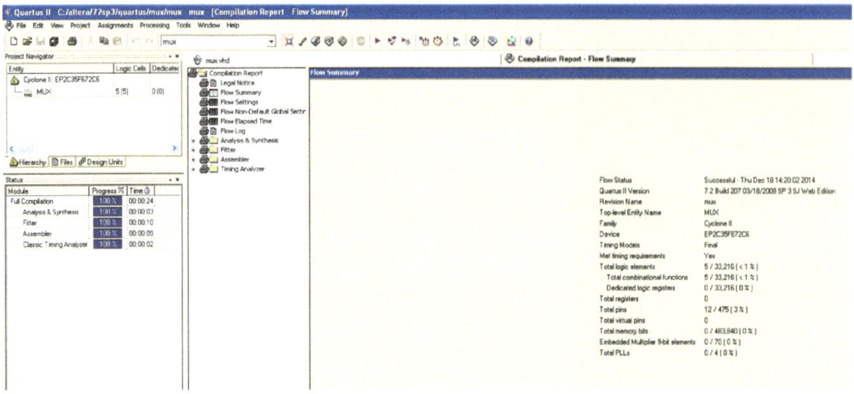

Fig. 3.5 Compilation report

If you want to see the design system at Netlist (gate) level, select tools > Netlist Viewer > RTL Viewer. After clicking on RTL Viewer, it will show that gate-level architecture as shown in Fig. 3.6.

Simulating the Design

1. After successfully compiling the designed system, one can go for simulating the waveforms.
2. Select **File** → then click on **New** → **Other Files** → **Vector Waveform Files**, and the window shown in Fig. 3.7 appears.

Save the waveform with .vwf extension in the same project directory.

3. One can specify ending time for the waveform to be simulated by clicking on **Edit** → **End time**, as shown in Fig. 3.8.
4. To see the full simulated output waveform, click on **View** → **Fit in Window**.
5. **To specify the input/output nodes, click on Edit** → **Insert Node or Bus**. Figure 3.9 shows the window of node finder.

Fig. 3.6 RTL of 8:1 multiplexer

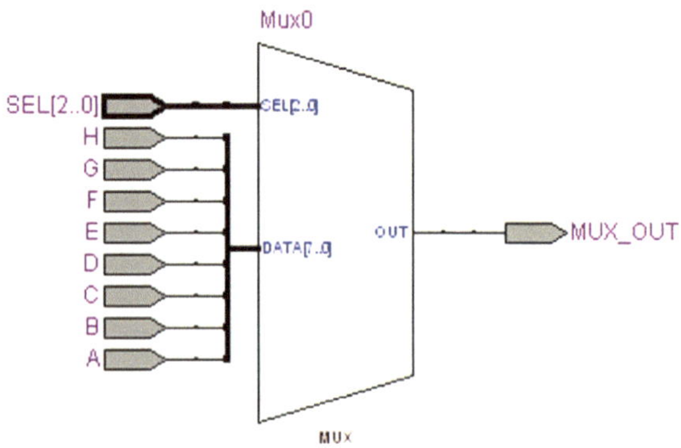

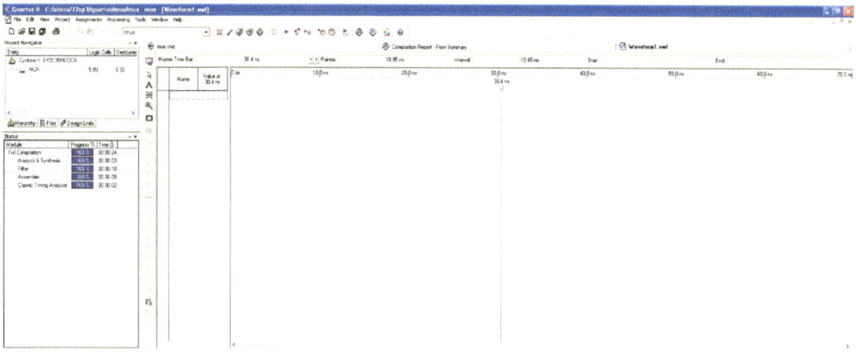

Fig. 3.7 Waveform editor window

Fig. 3.8 Assigning end time

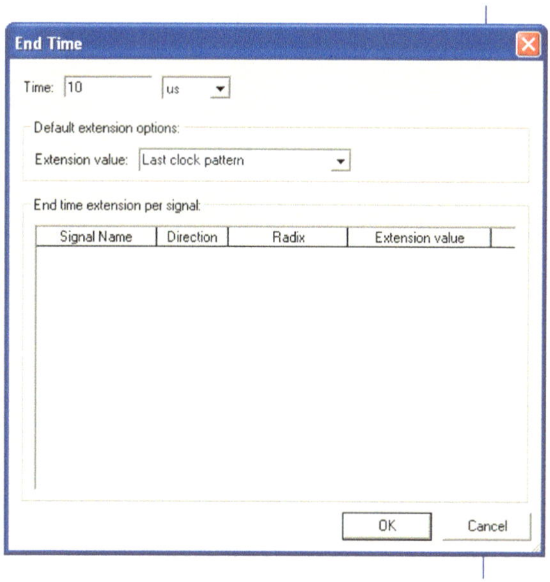

6. Select pins: all option under filter dialog box and then click on list option as shown in Fig. 3.9.
7. Clicking on list will give all the used pins in left pane. Select all the pins and click on > to move the pin to the right pane.
8. Now, you get all the pins in your waveform editor window as shown in Fig. 3.10.
9. Now assign the values to the input signal and select line of multiplexer as per the truth table of multiplexer. Assigned waveform window is shown in Fig. 3.11.

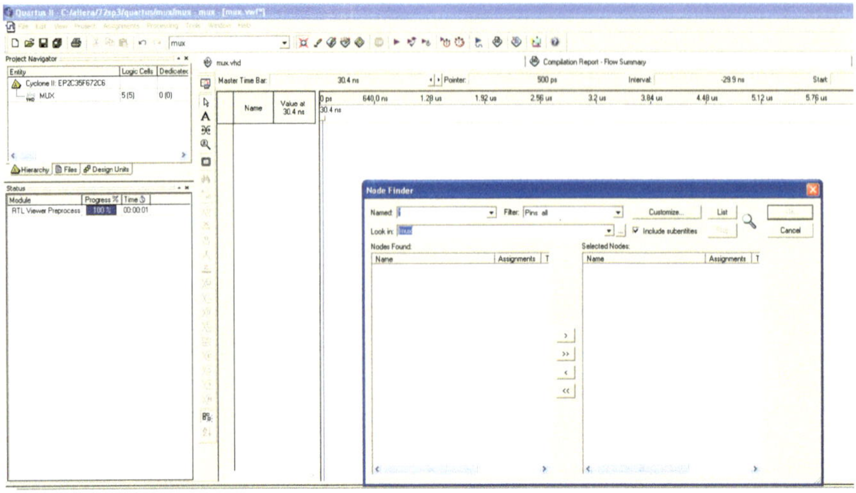

Fig. 3.9 Node finder window

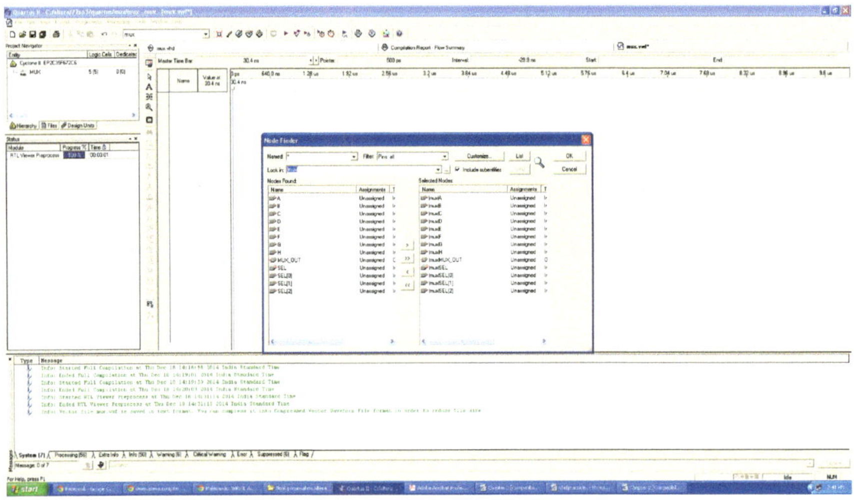

Fig. 3.10 Node finder with all pins list

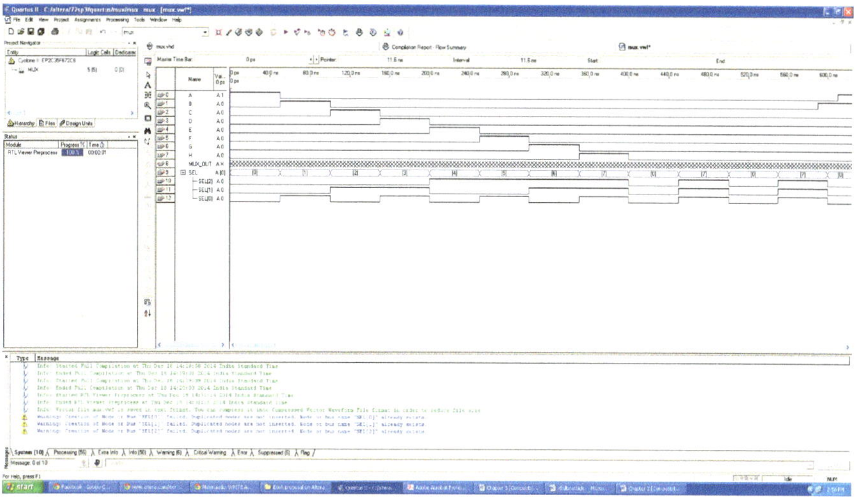

Fig. 3.11 Assigned waveform window

10. Quartus II simulator tool supports functional and timing simulation. Simulation mode is selected by clicking **Assignment → Settings → Simulator**.
11. Click on **Processing tab → click on Generate Functional Simulation Netlist**.
12. Then click on **Processing tab → select Start Simulation**, and the final simulated output of 3 inputs and gate is shown in Fig. 3.12.
13. Now, the next step is to download the design on target board and test the design.

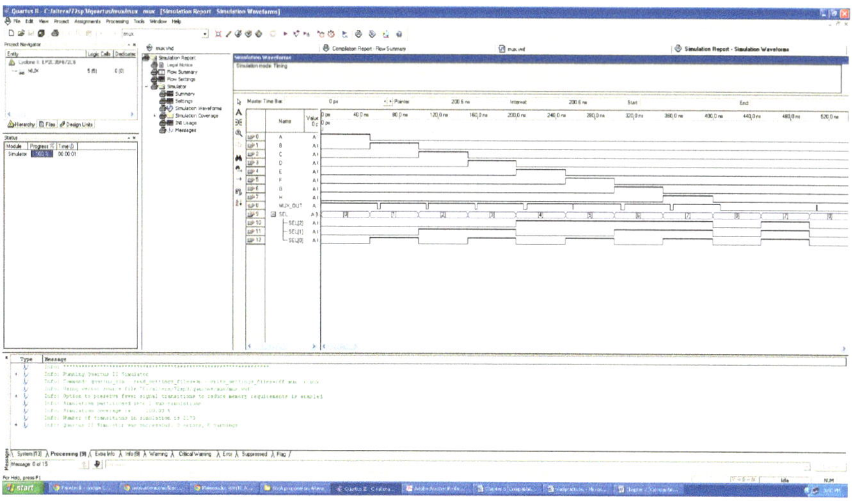

Fig. 3.12 Simulated waveform

Fig. 3.13 1:8 demultiplexer logic diagram

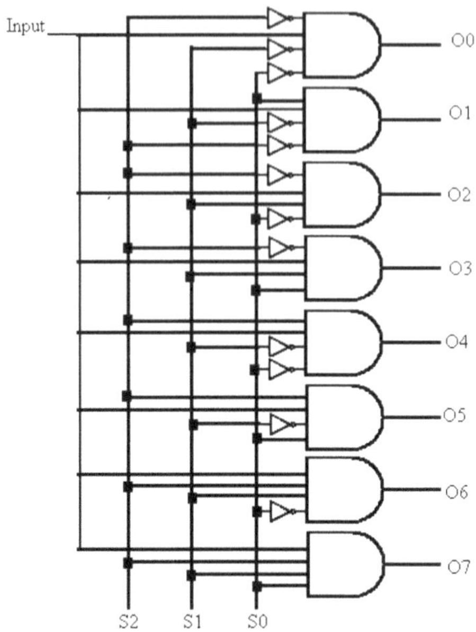

Demultiplexer

A device which takes single-input signal and selects one of the many output data lines that are connected to the single input is called demultiplexer. Sometimes, demultiplexer is also called as single-input, multiple-output switch. Demultiplexer transfers the same input data to multiple destinations; hence, it is also called as "distributor."

To design general-purpose logic, demultiplexers offer convenient solution. Demultiplexer is a combinational logic circuit which performs exactly the reverse operation what is done by multiplexer. It has single input, m select lines, and 2^m outputs. Here, one of the outputs will be selected based on the settings of select line to take the current state of input line. Figure 3.13 shows the 1:8 demultiplexer logic diagram.

1:8 Demultiplexer Truth Table

Select inputs			Output (D_{out})							
A	*B*	*C*	Dout (7)	Dout (6)	Dout (5)	Dout (4)	Dout (3)	Dout (2)	Dout (1)	Dout (0)
0	0	0	0	0	0	0	0	0	0	I
0	0	1	0	0	0	0	0	0	I	0
0	1	0	0	0	0	0	0	I	0	0
0	1	1	0	0	0	0	I	0	0	0

(continued)

(continued)

Select inputs			Output (D_{out})							
A	*B*	*C*	Dout (7)	Dout (6)	Dout (5)	Dout (4)	Dout (3)	Dout (2)	Dout (1)	Dout (0)
1	0	0	0	0	0	I	0	0	0	0
1	0	1	0	0	I	0	0	0	0	0
1	1	0	0	I	0	0	0	0	0	0
1	1	1	I	0	0	0	0	0	0	0

Procedure for Demux Implementation

Steps for creating the demultiplexer designs are same as that of multiplexer, and the design code for 1:8 demultiplexers is given below.

```
----------------------------------------------------------------------
-- Title      : demultiplexer
-- Design     : vhdl_test
-- Author     : Dr.J.S.Parab
----------------------------------------------------------------------


library IEEE;
use IEEE.STD_LOGIC_1164.all;

entity demuxp is
    port(
       dini : in STD_LOGIC;
       sel : in STD_LOGIC_VECTOR(2 downto 0);
       dout : out STD_LOGIC_VECTOR(7 downto 0)
       );
end demuxp;

architecture behavioral of demuxp is
begin

   dout <= (dini & "0000000") when (sel="000") else
         ('0' & dini & "000000") when (sel="001") else
         ("00" & dini & "00000") when (sel="010") else
         ("000" & dini & "0000") when (sel="011") else
         ("0000" & dini & "000") when (sel="100") else
         ("00000" & dini & "00") when (sel="101") else
         ("000000" & dini & '0') when (sel="110") else
         ("0000000" & dini) ;
   end behavioral;
```

Once the project is created, entire project is compiled, compilation report is generated as shown in Fig. 3.14, and the simulated output is shown in Fig. 3.15.

Now, the next step is to download the design of target board (DE2) and to test the design.

Flow Status	Successful - Wed Dec 24 10:52:08 2014
Quartus II Version	7.2 Build 207 03/18/2008 SP 3 SJ Web Edition
Revision Name	demux
Top-level Entity Name	demux
Family	Cyclone II
Device	EP2C35F672C6
Timing Models	Final
Met timing requirements	Yes
Total logic elements	8 / 33,216 (< 1 %)
Total combinational functions	8 / 33,216 (< 1 %)
Dedicated logic registers	0 / 33,216 (0 %)
Total registers	0
Total pins	12 / 475 (3 %)
Total virtual pins	0
Total memory bits	0 / 483,840 (0 %)
Embedded Multiplier 9-bit elements	0 / 70 (0 %)
Total PLLs	0 / 4 (0 %)

Fig. 3.14 Compilation report

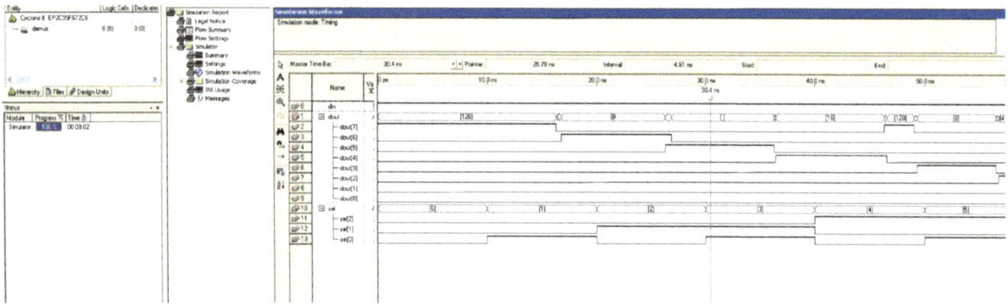

Fig. 3.15 Simulated output of 1:8 demultiplexers

3.2 Implementation of Encoder/Decoder and Priority Encoder

Encoder

The device which converts message information from one code format to another is called an encoder. The use of encoder in any design is to standardize and enhance speed and high level of security. An encoder has *M* input lines and *N* output lines. At a given time, only one input line is activated out of *M* input lines and it generates equivalent code on *N* output lines.

Octal-to-Binary Encoder (8:3)

Octal-to-binary encoder accepts 8 inputs and generates 3 outputs. At any given time, only one input line is active, i.e., its value is 1. The truth table of 8:3 encoder is given in Table 3.1. The logic diagram of 8:3 encoder is shown in Fig. 3.16.

For an 8:3 binary encoder having inputs as *I*0–*I*7, the outputs' logic expressions are as follows:

$$Y0 = I1 + I3 + I5 + I7$$
$$Y1 = I2 + I3 + I6 + I7$$
$$Y2 = I4 + I5 + I6 + I7$$

Table 3.1 Truth table of 8:3 encoder

*I*0	*I*1	*I*2	*I*3	*I*4	*I*5	*I*6	*I*7	Y_2	Y_1	Y_0
1	0	0	0	0	0	0	0	0	0	0
0	1	0	0	0	0	0	0	0	0	1
0	0	1	0	0	0	0	0	0	1	0
0	0	0	1	0	0	0	0	0	1	1
0	0	0	0	1	0	0	0	1	0	0
0	0	0	0	0	1	0	0	1	0	1
0	0	0	0	0	0	1	0	1	1	0
0	0	0	0	0	0	0	1	1	1	1

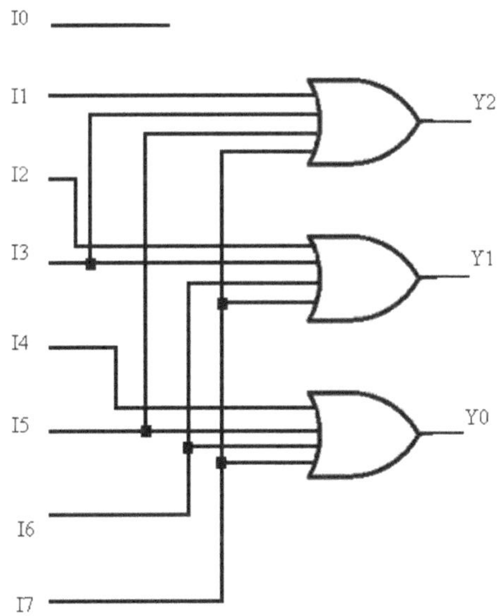

Fig. 3.16 8:3 encoder logic diagram

Implementation Procedure in Short

Step 1 Double click on Quartus—II 7.2.

Step 2 Go to file → New Project wizard and then click Next, and then, one has to give the working directory name, project name, and top-level entity name.

Step 3 Then click on next 2 times which will ask to specify the device (Cyclone II EP2C35F672C6).

Type your code for 8:3 encoder and save file with the same name as your entity.

```
---------------------------------------------------------------------------
-- Title      : Encoder
-- Design      : vhdl_test
-- Author      : Dr.J.S.Parab
---------------------------------------------------------------------------
library IEEE;
use IEEE.STD_LOGIC_1164.all;
entity encoder8 is
 port(
 Edin : in STD_LOGIC_VECTOR(7 downto 0);
       Edout : out STD_LOGIC_VECTOR(2 downto 0)
        );
end encoder8;
architecture encoder8:3 of encoder8 is
begin
Edout <= "000" when (Edin="10000000") else
       "001" when (Edin="01000000") else
       "010" when (Edin="00100000") else
       "011" when (Edin="00010000") else
       "100" when (Edin="00001000") else
       "101" when (Edin="00000100") else
       "110" when (Edin="00000010") else
       "111";

end encoder8:3;
```

Step 4 Next step is the assignment of FPGA pins.

Step 5 **Building and compilation of project.**

Compile it using tab Processing → start compilation.

After successful analysis and synthesis, compilation report is generated as shown in Fig. 3.17.

If you want to see the design system at Netlist (gate)-level, select tools > Netlist Viewer > RTL Viewer. After clicking on RTL Viewer, it will show that gate-level architecture as shown in Fig. 3.18.

Simulating the Design

Next step is to simulate the design. The simulated output is shown in Fig. 3.19.

Now, the next step is to download the design on target board (DE2) and to test the design.

Flow Status	Successful - Wed Aug 13 11:10:47 2014
Quartus II Version	7.2 Build 207 03/18/2008 SP 3 SJ Web Edition
Revision Name	encoder
Top-level Entity Name	encoder
Family	Cyclone II
Device	EP2C35F672C6
Timing Models	Final
Met timing requirements	Yes
Total logic elements	9 / 33,216 (< 1 %)
Total combinational functions	9 / 33,216 (< 1 %)
Dedicated logic registers	0 / 33,216 (0 %)
Total registers	0
Total pins	11 / 475 (2 %)
Total virtual pins	0
Total memory bits	0 / 483,840 (0 %)
Embedded Multiplier 9-bit elements	0 / 70 (0 %)
Total PLLs	0 / 4 (0 %)

Fig. 3.17 Compilation report

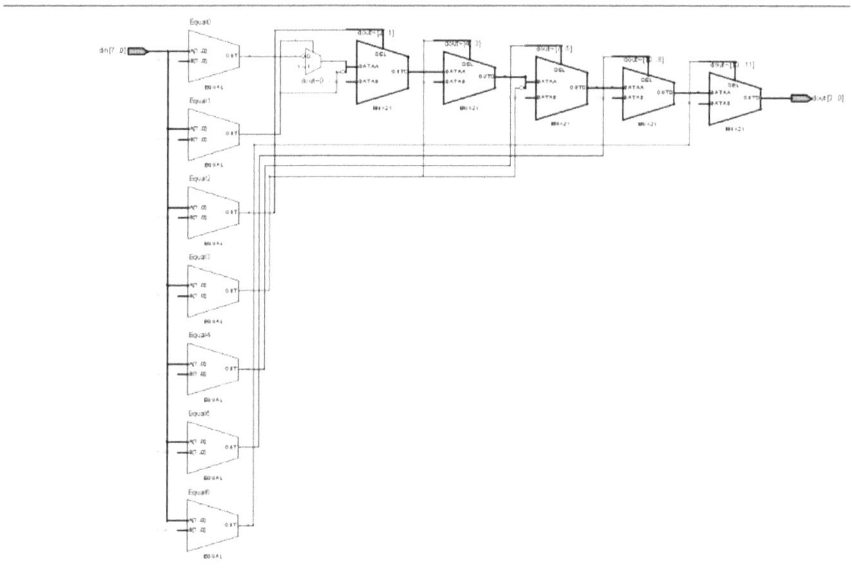

Fig. 3.18 RTL of 8:3 encoder

Decoder

A decoder performs exactly the reverse operation that what is performed by encoder to recover the original information.

In digital electronics, decoding is required in applications involving multiplexing of data, interfacing seven-segment display, and addressing decoding for memory interfaced to microprocessor or microcontroller.

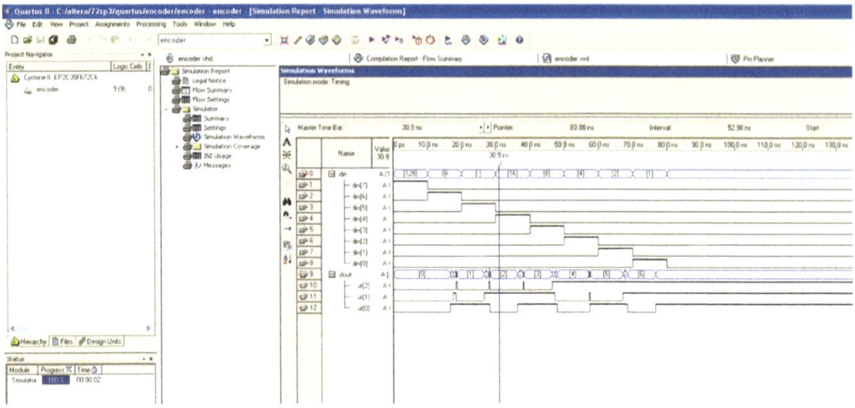

Fig. 3.19 Simulated output waveform

Procedure

Steps for creating the decoder design are same as that of multiplexer, and the design code for 3:8 decoder is given below.

```
----------------------------------------------------------------------------
-- Title      : decoder
-- Design     : vhdl_test
-- Author     : Dr.J.S.Parab
----------------------------------------------------------------------------

library IEEE;
use IEEE.STD_LOGIC_1164.all;

entity decoder8 is
    port(
        Ddin : in STD_LOGIC_VECTOR(2 downto 0);
        Ddout : out STD_LOGIC_VECTOR(7 downto 0)
        );
end decoder8;

architecture decoder3:8 of decoder8 is
begin

  Ddout <= ("10000000") when (Ddin="000") else
        ("01000000") when (Ddin="001") else
        ("00100000") when (Ddin="010") else
        ("00010000") when (Ddin="011") else
        ("00001000")  when (Ddin="100") e lse
        ("00000100") when (Ddin="101") e lse
        ("00000010") when (Ddin="110") el se
        ("00000001") ;

end decoder3:8;
```

Compilation Report

Once the project is created, entire project is compiled and compilation report is generated as shown below.

Flow Status	Successful - Thu Jul 31 15:28:24 2014
Quartus II Version	7.2 Build 207 03/18/2008 SP 3 SJ Web Edition
Revision Name	decoder
Top-level Entity Name	decoder
Family	Cyclone II
Device	EP2C35F672C6
Timing Models	Final
Met timing requirements	Yes
Total logic elements	8 / 33,216 (< 1 %)
Total combinational functions	8 / 33,216 (< 1 %)
Dedicated logic registers	0 / 33,216 (0 %)
Total registers	0
Total pins	11 / 475 (2 %)
Total virtual pins	0
Total memory bits	0 / 483,840 (0 %)
Embedded Multiplier 9-bit elements	0 / 70 (0 %)
Total PLLs	0 / 4 (0 %)

Decoder Simulated Output

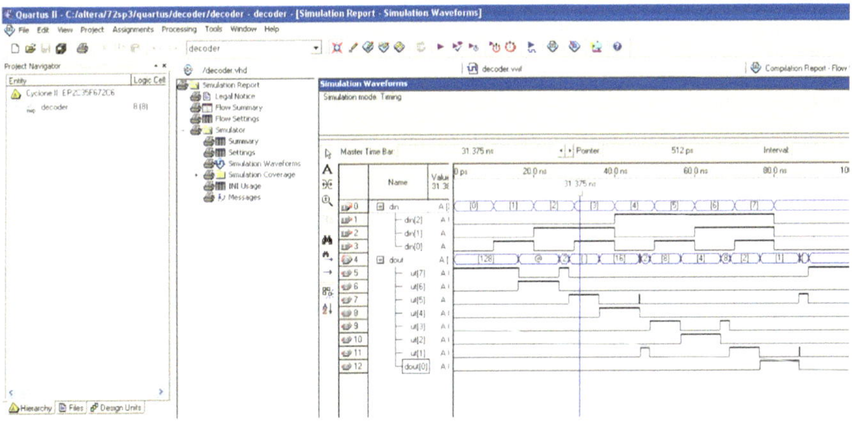

The RTL schematic of 3:8 decoder is shown below

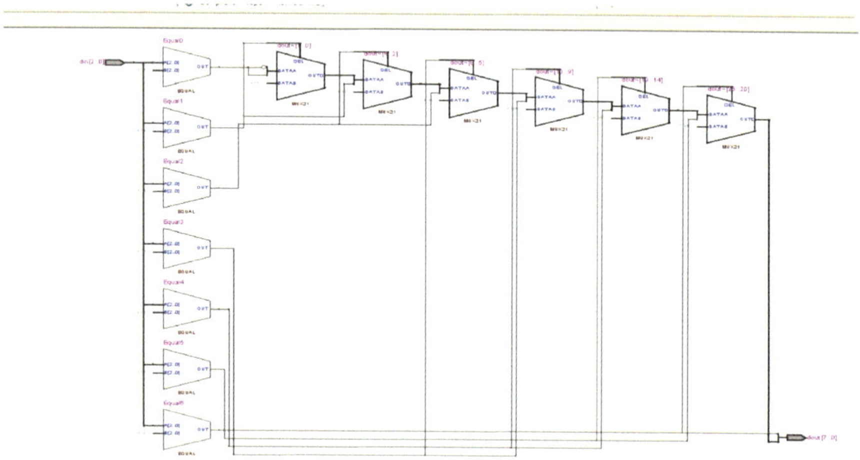

Now the next step is to download the design on target board (DE2) and to test the design.

Priority Encoder

A circuit which compresses several binary inputs into a few number of outputs is called priority encoder. Sometimes, priority encoder is used to manage interrupt requests by considering the priority of interrupt. If several inputs are active at same time, the input which has the highest priority will be selected.

```
---------------------------------------------------------------------------
-- Title      : priority encoder
-- Design     : vhdl_test
-- Author     : Dr.J.S.Parab
---------------------------------------------------------------------------
  LIBRARY ieee;
USE ieee.std_logic_1164.ALL;
entity prioencd is
port(P : in bit_vector(7 downto 0);            --inputs to be prioritized
M : out bit_vector(2 downto 0));                 --encoded output
end prioencd;
```

```
architecture priority of prioencd is
begin
process(P)
begin
M <= "000";
if P(7) = '1' then
M <= "111";
elseif P(6) = '1' then
M <= "110";
elseif P(5) = '1' then
M <= "101";
elseif P(4) = '1' then
M <= "100";
elseif P(3) = '1' then
M <= "011";
elseif P(2) = '1' then
M <= "010";
elseif P(1) = '1' then
M <= "001";
elseif P(0) = '1' then
M <= "000";
end if;
end process;
end priority;
```

Compilation Report

Flow Status	Successful - Wed Aug 13 12:22:07 2014
Quartus II Version	7.2 Build 207 03/18/2008 SP 3 SJ Web Edition
Revision Name	priorityencd
Top-level Entity Name	priorityencd
Family	Cyclone II
Device	EP2C35F672C6
Timing Models	Final
Met timing requirements	Yes
Total logic elements	5 / 33,216 (< 1 %)
Total combinational functions	5 / 33,216 (< 1 %)
Dedicated logic registers	0 / 33,216 (0 %)
Total registers	0
Total pins	11 / 475 (2 %)
Total virtual pins	0
Total memory bits	0 / 483,840 (0 %)
Embedded Multiplier 9-bit elements	0 / 70 (0 %)
Total PLLs	0 / 4 (0 %)

Simulated Output

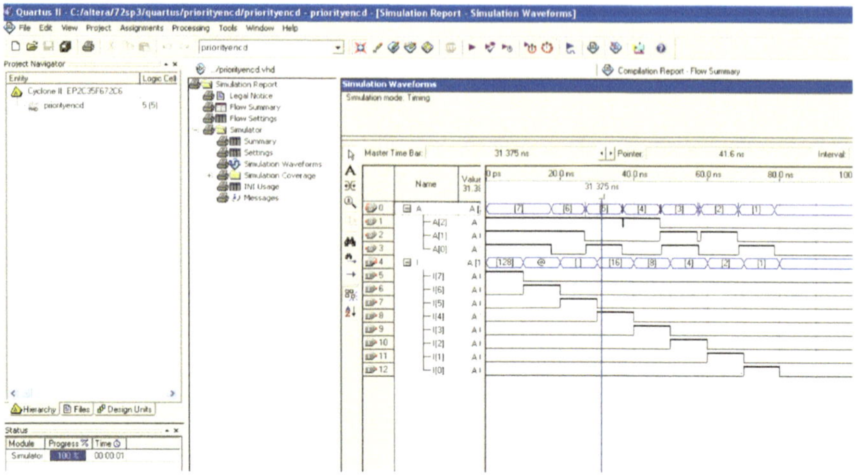

Priority RTL Schematics

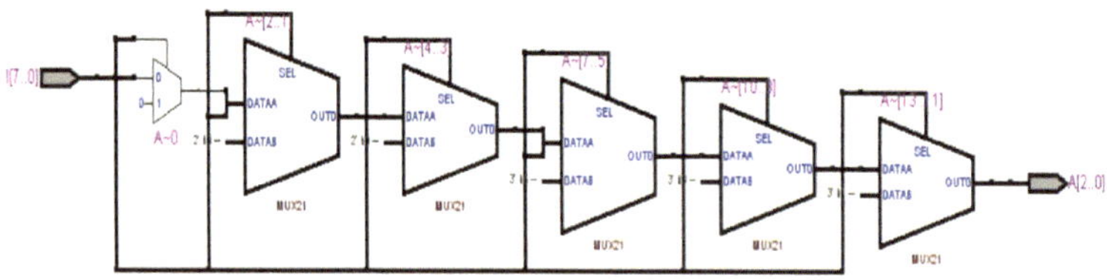

3.3 Universal Shift Register

The shift register is a logic circuit used to store the information message in binary form. This device loads new data available on its inputs pin, and then, that data gets shifts to the output on each clock cycle, that's why the name "shift register."

A universal shift register supports three different modes of transferring data: (1) It accepts data in parallel and transmits data in parallel (PIPO). (2) Data comes in serial and outputted in serial (SISO) may be through left shifts or right shifts. (3) Universal register can load data in series and then output data in parallel (SIPO).

Implementation Procedure in Short

Step 1 Double click on Quartus—II 7.2.

Step 2 Go to file → New Project wizard and then click Next, and then, one has to give the working directory name, project name, and top-level entity name.

Step 3 Then click on next 2 times which will ask to specify the device (Cyclone II EP2C35F672C6).
Type your code for universal shift register and save file with the same name as your entity (i.e., shift register).

```
--------------------------------------------------------------------------
-- Title     : universal shift register
-- Design    : vhdl_test
-- Author    : Dr.J.S.Parab
--------------------------------------------------------------------------
library ieee;

use ieee.std_logic_1164.all;

entity shftregister is
port(C, SInp, left_right,SLD :in std_logic;
    D : in std_logic_vector(7 downto 0);
SOut:out std_logic;
 POut : out std_logic_vector(7 downto 0));
end shftregister;

architecture unisal of shftregister is
 signal temp: std_logic_vector(7 downto 0);
 begin
 process (C)
 begin
 if (C'event and C='1') then
if (SLD='1') then
 tmp <= D;
else
 if (left_right='0') then
 temp <= (temp(6 downto 0) & SI);
SOut <=temp(7);
 else
 temp <= (SInp & temp(7 downto 1));
 end if;
 end if;
end if;
 end process;

 POut <= temp;
end unisal;
```

Step 4 Next step is the assignment of FPGA pins.
Step 5 **Building and compilation of project**.

Compile it using tab Processing → start compilation.

After successful analysis and synthesis, compilation report is generated as shown below.

Compilation Report

Flow Status	Successful - Fri Jan 02 16:38:10 2015
Quartus II Version	7.2 Build 207 03/18/2008 SP 3 SJ Web Edition
Revision Name	shiftregister
Top-level Entity Name	shiftregister
Family	Cyclone II
Device	EP2C35F672C6
Timing Models	Final
Met timing requirements	Yes
Total logic elements	9 / 33,216 (< 1 %)
Total combinational functions	9 / 33,216 (< 1 %)
Dedicated logic registers	9 / 33,216 (< 1 %)
Total registers	9
Total pins	21 / 475 (4 %)
Total virtual pins	0
Total memory bits	0 / 483,840 (0 %)
Embedded Multiplier 9-bit elements	0 / 70 (0 %)
Total PLLs	0 / 4 (0 %)

After clicking on RTL Viewer, it will show that gate-level architecture as shown below.

RTL Viewer

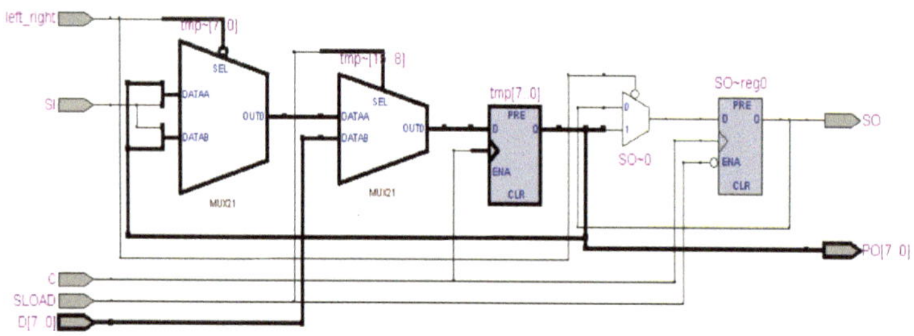

Next step is to simulate the design, and the simulated output is shown below.

Simulated Waveform Output:

PIPO

Here, SLOAD line = "1" which indicates parallel out.

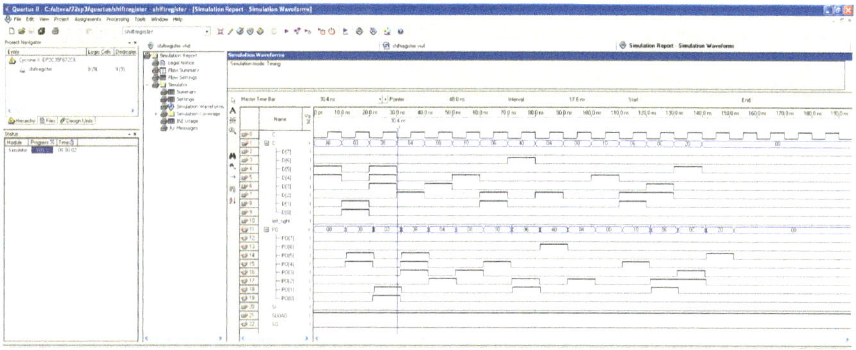

SIPO

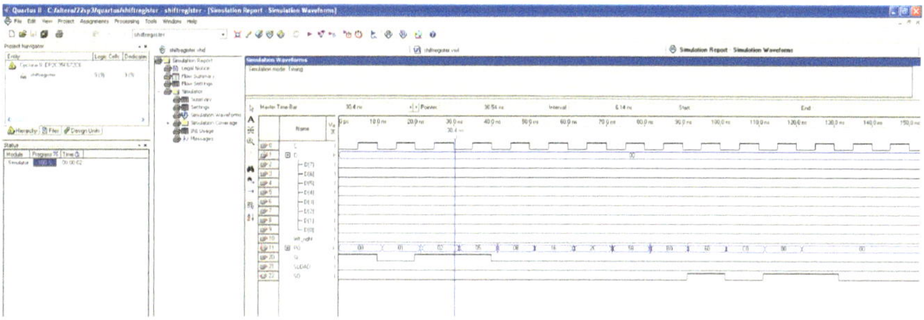

SIPO Left Shift and SLOAD = "0"

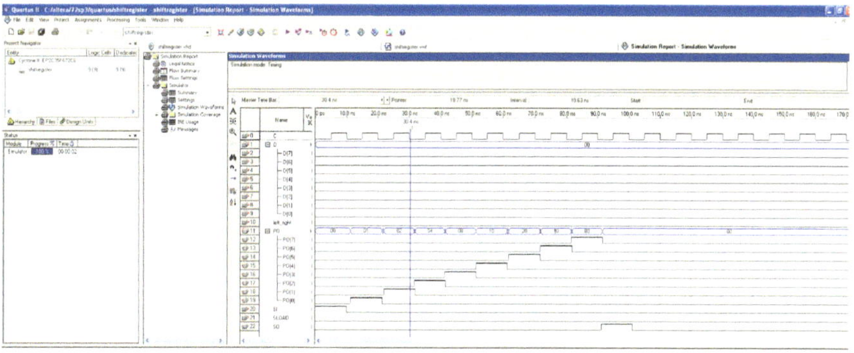

SIPO Right Shift and SLOAD = "0"

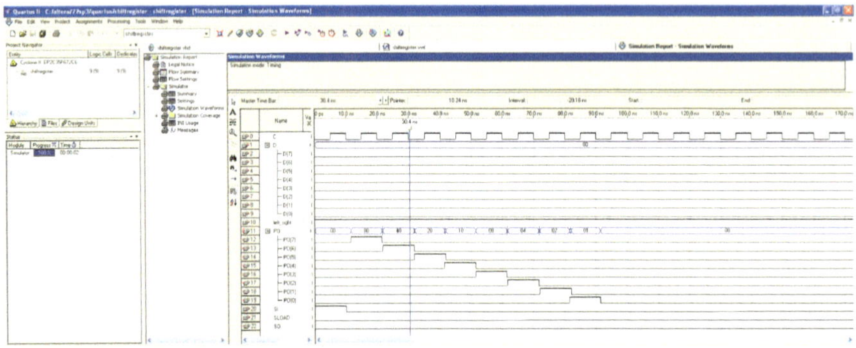

Now, the next step is to download the design on target board (DE2) and to test the design.

3.4 4-Bit Counter

Counter is a device which is used to count the occurrence of particular event with respect to the clock signal. There are basically two types of counters, i.e., asynchronous and synchronous counter.

Bidirectional Counter

There are universal types of counter which counts in both directions, i.e., up or down, based on the state of their mode control input. These bidirectional counters are capable of counting in both directions, and their mode can be reversed at any point within their count sequence by changing the control line.

Synchronous 4-bit Up/Down Counter:

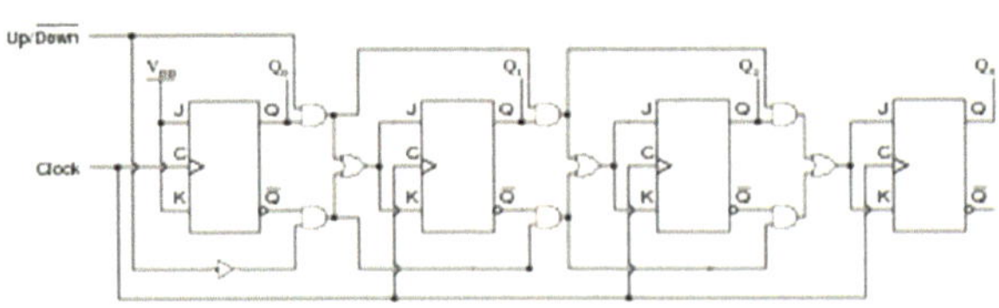

Procedure in Short

Step 1 Double click on Quartus—II 7.2.

Step 2 Go to file → New Project wizard and then click Next, and then, one has to give the working directory name, project name, and top-level entity name.

Step 3 Then click on next 2 times which will ask to specify the device (Cyclone II EP2C35F672C6).
Type your code for 4-bit counter and save file with the same name as your entity (i.e., counter).

```
---------------------------------------------------------------------------
-- Title      : 4 bit counter
-- Design     : vhdl_test
-- Author     : Dr.J.S.Parab
---------------------------------------------------------------------------
          library ieee;
          use ieee.std_logic_1164.all;
          use ieee.std_logic_unsigned.all;

          entity counter4 is
               port(Clk, CLEAR, up_down : in std_logic;
                     count : out std_logic_vector(3 downto 0));
           end counter4;

          architecture updwn of counter4 is
          signal temp: std_logic_vector(3 downto 0);

           begin
           process (Clk, CLEAR)
           begin
           if (CLEAR='1') then
           temp <= "0000";
           elseif (Clk 'event and Clk='1') then
           if (up_down='1') then
           temp <= temp + 1;
           else
           temp <= temp - 1;
           end if;
           end if;
           end process;
          count <= temp;
end updwn;
```

Step 4 Next step is the assignment of FPGA pins.
Step 5 **Building and compilation of project**.

Compile it using tab Processing → start compilation.

After successful analysis and synthesis, compilation report is generated as shown below.

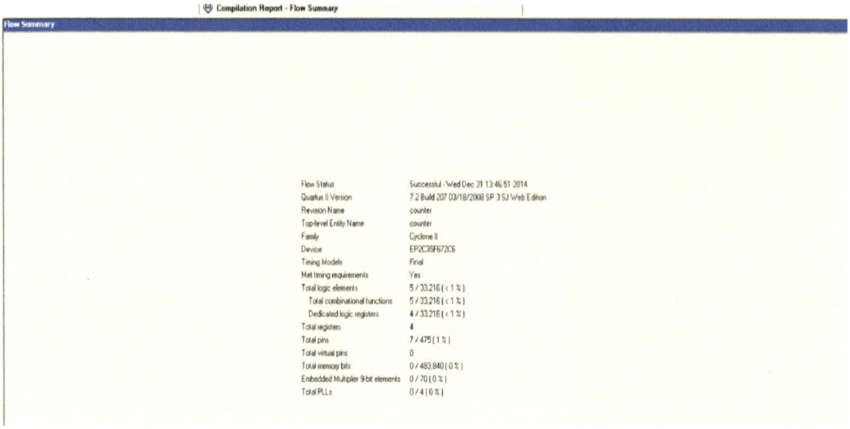

After clicking on RTL Viewer, it will show that gate-level architecture as shown below.

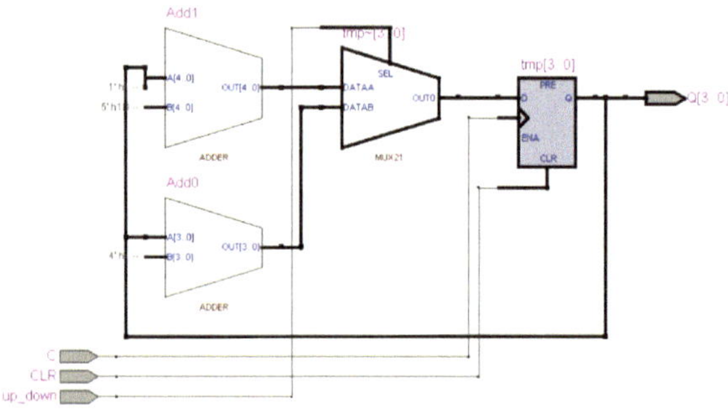

Next step is to simulate the design, and the simulated output is shown below.

Simulated Waveform Output

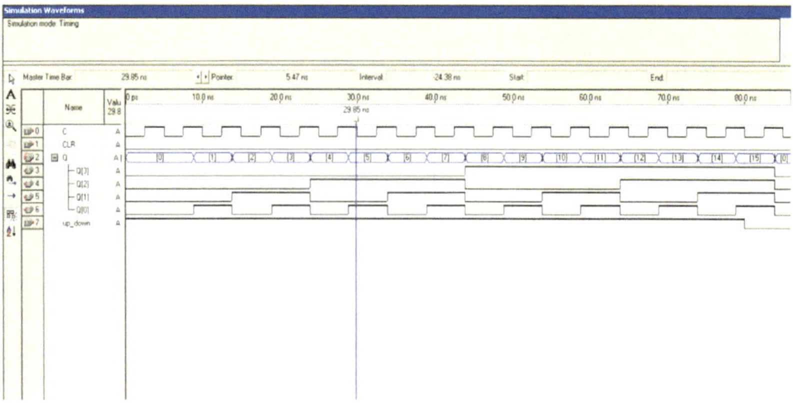

Now, the next step is to download the design on target board (DE2) and to test the design.

3.5 Implementation of Memory

There are various types of memories available. Here, we will consider only random-access memory (RAM). RAM is volatile in nature and used to store the data temporarily.

The VHDL code we have implemented includes reading and writing to RAM.

Address bus (as $2^8 = 256$).

Hence, each location can store 8 bits (i.e., 1 byte each).

Procedure in Short

Step 1 Double click on Quartus—II 7.2.

Step 2 Go to file → New Project wizard and then click Next, and then, one has to give the working directory name, project name, and top-level entity name.

Step 3 Then click on next 2 times which will ask to specify the device (Cyclone II EP2C35F672C6).

Type your code for memory and save file with the same name as your entity (i.e., memory).

```
-------------------------------------------------------------------------------
-- Title      : memory
-- Design      : vhdl_test
-- Author      : Dr.J.S.Parab
-------------------------------------------------------------------------------

library IEEE;

use IEEE.STD_LOGIC_1164.ALL;

use IEEE.STD_LOGIC_ARITH.ALL;

use IEEE.STD_LOGIC_UNSIGNED.ALL;

entity memory8 is
Port ( Clk : in std_logic;
wrte : in std_logic;
wradd : in std_logic_vector(6 downto 0);
rdadd : in std_logic_vector(6 downto 0);
datainp : in std_logic_vector(7 downto 0);
dataout : out std_logic_vector(7 downto 0));
end memory;
```

```
architecture RAM8 of Memory8 is
type ram is array(127 downto 0) of std_logic_vector(7 downto 0);
signal ram1_1 : ram;
signal rd_add : std_logic_vector(6 downto 0);

begin
process(Clk, wrte)
begin
if Clk'event and Clk = '1' then
if wrte = '1' then
ram1_1(conv_integer(wradd)) <= datainp;
end if;
rd_add <= rdadd;
end if;
end process;

dataout <= ram1_1(conv_integer(rd_add));
end RAM8;
```

Step 4 Next step is the assignment of FPGA pins.

Building and Compilation of Project

Compile it using tab Processing → start compilation.

After successful analysis and synthesis, compilation report is generated as shown below.

Compilation Report

Flow Status	Successful - Thu Jan 01 16:12:51 2015
Quartus II Version	7.2 Build 207 03/18/2008 SP 3 SJ Web Edition
Revision Name	memory
Top-level Entity Name	memory
Family	Cyclone II
Device	EP2C35F672C6
Timing Models	Final
Met timing requirements	Yes
Total logic elements	24 / 33,216 (< 1 %)
Total combinational functions	13 / 33,216 (< 1 %)
Dedicated logic registers	23 / 33,216 (< 1 %)
Total registers	23
Total pins	32 / 475 (7 %)
Total virtual pins	0
Total memory bits	1,024 / 483,840 (< 1 %)
Embedded Multiplier 9-bit elements	0 / 70 (0 %)
Total PLLs	0 / 4 (0 %)

After clicking on RTL Viewer, it will show that gate-level architecture as shown below.

RTL Viewer

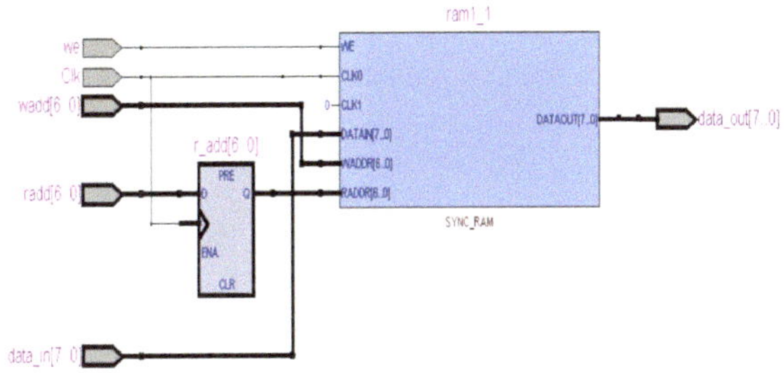

Next step is to simulate the design, and the simulated output is shown below.

Simulated Waveform Output

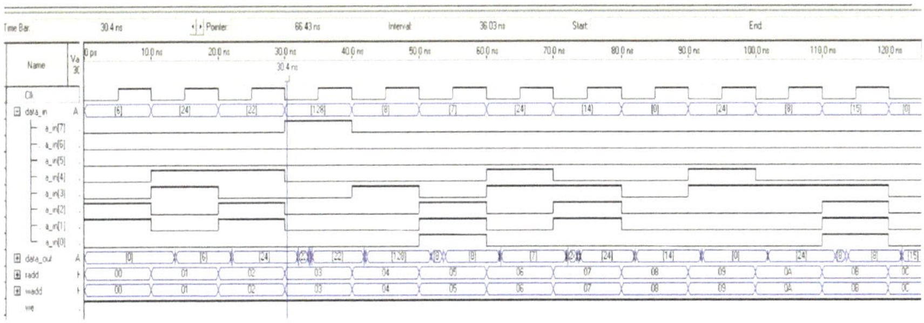

Now, the next step is to download the design on target board (DE2) and to test the design.

3.6 Traffic Light Controller

Traffic light controllers are installed mostly on the road junctions to solve the problem of traffic congestion and that will ease the smooth flow of traffic.

Figure 3.20 shows the set of traffic LED lights, and these lights are installed at junction with one rod going north–south and the other going east–west.

Here, we have implemented traffic light controller having red, yellow, and green LEDs using state machine approach with six states as shown in Table 3.2. Figure 3.21 shows the state diagram of traffic light controller.

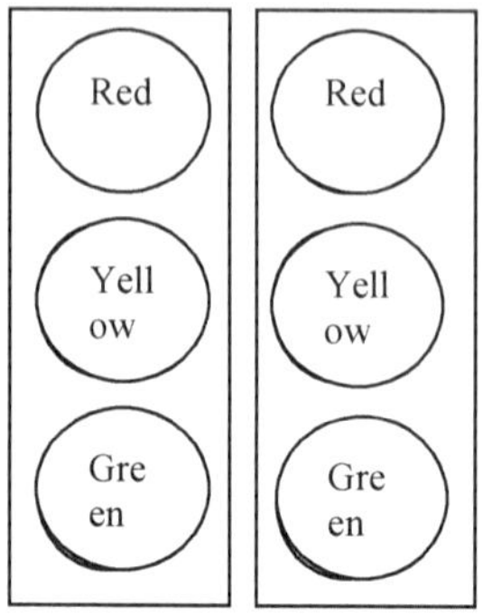

Fig. 3.20 Set of traffic lights

Table 3.2 Six states of TLC controller

State	*North–South LEDs*	*East–West LEDs*	*Delay in seconds*
0	Green	Red	5
1	Yellow	Red	1
2	Red	Red	1
3	Red	Green	5
4	Red	Yellow	1
5	Red	Red	1

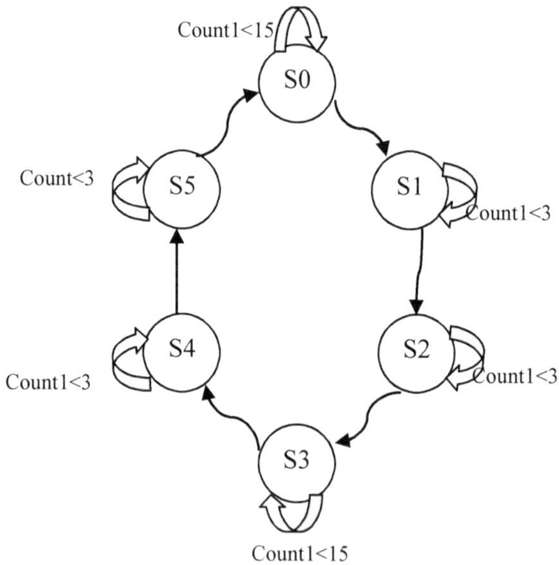

Fig. 3.21 State machine diagram

Implementation Procedure in Short

Step 1 Double click on Quartus—II 7.2.

Step 2 Go to file → New Project wizard and then click Next, and then, one has to give the working directory name, project name, and top-level entity name.

Step 3 Then click on next 2 times which will ask to specify the device (Cyclone II EP2C35F672C6).

Type your code for traffic light controller and save file with the same name as your entity (i.e., traffic light).

```
--------------------------------------------------------------------------------
-- Title       : Traffic lights controller
-- Design      : vhdl_test
-- Author      : Dr.J.S.Parab
Note:  Traffic lights controller based on state machine

--------------------------------------------------------------------------------
library IEEE;
use IEEE.STD_LOGIC_1164.all;
use IEEE.STD_LOGIC_unsigned.all;
entity TLC is
 port (clk: in STD_LOGIC;
 clr: in STD_LOGIC;
 ledlights: out STD_LOGIC_VECTOR(5 downto 0));
end TLC;
architecture trafficlight of TLC is
type stateTC_type is (s0, s1, s2, s3, s4, s5);
signal stateTC: state_type;
signal count1: STD_LOGIC_VECTOR(3 downto 0);
constant SECfive: STD_LOGIC_VECTOR(3 downto 0) := "1111";
constant SECone: STD_LOGIC_VECTOR(3 downto 0) := "0011";
begin
process(clk, clr)
begin
 if clr = '1' then
 stateTC <= s0;
 count1 <= X"0";
 elseif clock'event and clock = '1' then
 case stateTC is
 when s0 =>
 if count1 < SECFive then
 state <= s0;
 count1 <= count + 1;
 else
 state <= s1;
 count1 <= X"0";
 end if;
 when s1 =>
```

```
if count1 < SECOne then
state <= s1;
count1 <= count1 + 1;
else
state <= s2;
count1 <= X"0";
end if;
when s2 =>
if count1 < SECOne then
state <= s3;
count1 <= count1 + 1;
else
state <= s3;
count1 <= X"0";
end if;
when s3 =>
if count1 < SECFive then
state <= s3;
count 1<= count1 + 1;
else
state <= s4;
count1 <= X"0";
end if;
when s4 =>
if count1 < SEC1 then
state <= s4;
count1 <= count + 1;
else
state <= s5;
count1 <= X"0";
end if;
when s5 =>
if count1 < SEC1 then
state <= s5;
count1 <= count1 + 1;
else
state <= s0;
count1 <= X"0";
end if;
when others =>
state <= s0;
end case;
end if;
end process;
C2: process(stateTC)
begin
```

```
case stateTC is
 when s0 => ledlights <= "100001";
 when s1 => ledlights <= "100010";
 when s2 => ledlights <= "100100";
 when s3 => ledlights <= "001100";
 when s4 => ledlights <= "010100";
 when s5 => ledlights <= "100100";
 when others => ledlights <= "100001";
end case;
end process;
end trafficlight;
```

Step 4 Next step is the assignment of FPGA pins.

Step 5 **Building and compilation of project.**

Compile it using tab Processing → start compilation.

After successful analysis and synthesis, compilation report is generated as shown below.

Flow Summary	
Flow Status	Successful - Wed Dec 31 11:35:17 2014
Quartus II Version	7.2 Build 207 03/18/2008 SP 3 SJ Web Edition
Revision Name	trafficlight
Top-level Entity Name	trafficlight
Family	Cyclone II
Device	EP2C35F672C6
Timing Models	Final
Met timing requirements	Yes
Total logic elements	18 / 33,216 (< 1 %)
Total combinational functions	18 / 33,216 (< 1 %)
Dedicated logic registers	10 / 33,216 (< 1 %)
Total registers	10
Total pins	8 / 475 (2 %)
Total virtual pins	0
Total memory bits	0 / 483,840 (0 %)
Embedded Multiplier 9-bit elements	0 / 70 (0 %)
Total PLLs	0 / 4 (0 %)

After clicking on RTL Viewer, it will show that gate-level architecture as shown below.

RTL Viewer

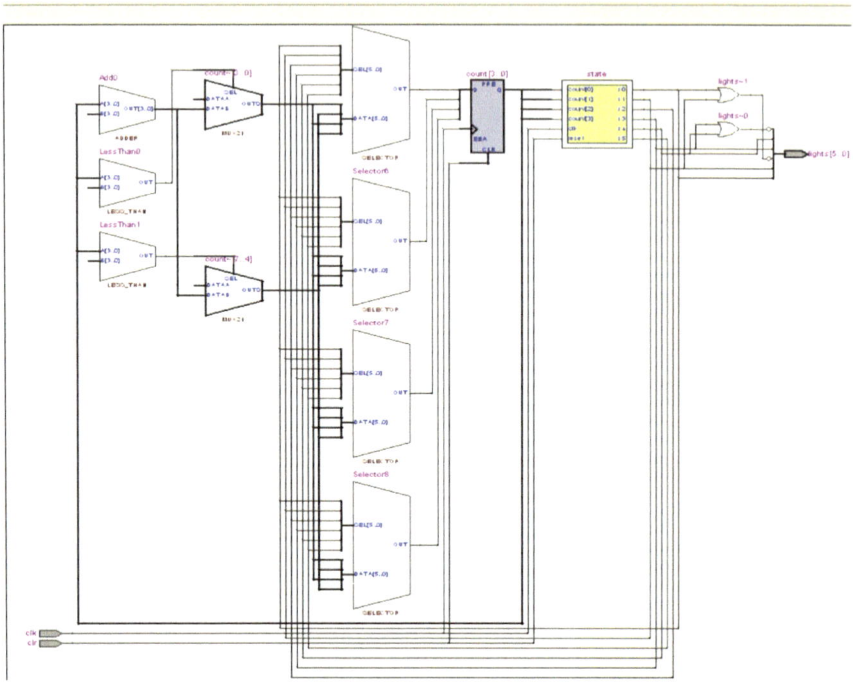

Next step is to simulate the design, and the simulated output is shown below.

Simulated Waveform Output

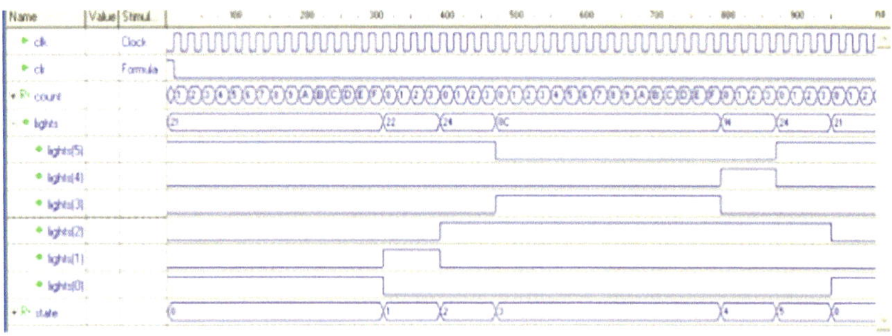

Now, the next step is to download the design on target board (DE2) and to test the design.

Chapter 4
Building Embedded Systems Using Soft IP Cores

Contents

Abstract This chapter makes reader aware of embedded soft core processors, their concepts, comparisons of various soft cores from various FPGA manufactures, etc. Here, we have emphasized on Altera Nios II soft core processor. The soft core nature of the Nios II processor lets the system designer specify and generate a custom Nios II core, tailored for his or her specific application requirements. System designers can extend the Nios II basic functionality by adding a predefined memory management unit or defining custom instructions and custom peripherals. Altera's Nios® II processor, the world's most versatile processor, according to Gartner Research, is the most widely used soft processor in the FPGA industry. Design development flow of Nios II System is also depicted in pictorial form which is self-explanatory for the reader.

Keywords Soft processor · Soft IP · Nios II processor

What is Intellectual Property (IP)?

IP is law protected component, e.g., patents, copyright, and trademarks. IP allows people to earn lot of recognition and also economic benefit with their invention or creation.

J.S. Parab et al., *Hands-on Experience with Altera FPGA Development Boards*,
https://doi.org/10.1007/978-81-322-3769-3_4

IP core in technology domain, IP is copyright property of individual part which can be a reusable unit or chip layout. ASIC chip and FPGA logic designs make use of IP cores so as to make the design less complex. There are various reasons to protect and publicize the IP.

- To create and invent new designs in technology domain.
- Further innovation by putting more resources.
- Protecting IP helps to enhances economic growth and creates new jobs.

4.1 Concept of Soft IPs

European Patent Forum was the first to use the term “Soft IP” in 2007. Soft IP is synthesizable form licensed product and can be available to the different parties by purchasing the licensed copy. Soft IPs are design rights, copyrights, or trademarks but hard IPs are patents.

In digital design industry, IP cores can be of Gate Netlist form or synthesizable form. IP with Netlist form is nothing but Boolean representation, i.e., with gates and standards cells similar to assembly listing of high-level program. One cannot apply reverse engineering on Netlist-based IP cores. Synthesizable IP versions are available in HDL languages which allow customer to ulter the design at functional level.

Soft IP provides customer with lot of design flexibility. It also offers better predictive nature in terms of timing performance. In electronic design industry, IP cores play major role in SoCs design.

4.2 Soft Core Processors for Embedded Systems

In today’s era, everybody talks about an Embedded Systems, which is nothing but a hardware and software combination to achieve desired task. Designing an embedded-based product is a challenging task as it has to meet constraints on area usage, size, power consumption, and also time to market. Over the years, complexity of the embedded system design has increased even if small change in the design requires the designing from scratch which leads to lot of time consumption. Hence, predesigned and tested IP cores are the alternative to solve the above problem. There are many advantages of using soft IPs in embedded design.

- They are flexible in nature and can easily be customizable.
- Less chances of getting obsolete since it can be synthesizable for any target device.
- Soft IPs are described in HDL; hence, it is easier to understand the entire design.

Several FPGA manufactures in market provides their soft core IPs for various blocks to attract the designers toward their products.

Altera has developed various IP cores so as to increase the usage and attract the better market for their development boards. These IPs can be easily incorporated in the design by using Quartus II system integration tool, Qsys. Altera also provides number of other IP cores, which are available as Altera IP MegaStore.

4.3 A Survey of Soft Core Processors

Here, we will discuss in brief the soft cores provided by various vendors.

4.3.1 Commercial Cores and Tools

Altera and Xilinx corporation have developed leading soft core processor such as NIOS II, MicroBlaze, and PicoBlaze. We will briefly discuss the some important features of various soft core processors.

NIOS II by Altera Corporation
Altera corporation has marked its footprint in market as leading vendor of CPLDs and FPGAs. They offer various range of FPGA variants such as Stratix, Stratix II, and Cyclone. In any embedded system design, the NIOS II processor can be instantiated just by simple selection process in SOPC Builder.

The NIOS II soft core processor is a RISC processor core and supports Harvard memory architecture. Nios II has 32-bit ISA, 32 general purpose registers and single instruction 32 $\times$ 32 multiply and divide. Nios II has three versions: economy, standard, and fast core. Each version comes with variation pipeline stages, instructions/data cache memories and their performance also varies.

MicroBlaze and PicoBlaze by Xilinx Incorporated
Spartan and Virtex FPGA are the devices of Xilinx Incorporated. In addition, they offer MicroBlaze and PicoBlaze soft core processor which is 32 bit. It is based on Harvard RISC architecture.

The MicroBlaze soft core processor developed is targeted on Virtex and Spartan families of FPGAs only. Xilinx also provides set of other IP cores which are required to design the embedded system. Xilinx also supplies lower version of soft core which is PicoBlaze soft core, which is 8-bit Microcontroller targeted on low-end FPGA like Spartan-3, Virtex-II, and Virtex-II Pro families of FPGAs. The PicoBlaze mostly used for simple data processing applications.

Soft Cores from Other Vendors
Tensilica Inc. offers a number of low cost, power-optimized soft IP processing cores for embedded systems design. These cores are mostly used for DSP application.

Tensilica's Xtensa Series processors has "configurable" feature which allows designer to tune the processor as per his intended application by varying the pre-defined parameters.

4.3.2 *Open-Source Cores*

There are large number of open-source cores are freely available. These cores are mostly used by academia for research and development of their embedded system-based product. Earlier, Altera has come out with UT NIOS open-core processor. Open SPARC from Sun Microsystems, *LEON by Gaisler Research,* and OpenRISC 1200 soft core processors that are available in open-source community.

4.3.3 *Comparison of Soft Core Processors*

Table 4.1 below shows a quick and best comparison about the various soft core processors from different vendors.

Soft core processors such as NIOS II and MicroBlaze designed for system to be implemented on FPGA. In contrast, the other three cores are not meant for specific target technology.

4.4 Soft Processor Cores of Altera

The most popular and widely used processor in FPGA industry is Altera Nios II as per Gartner Research. The processor can be implemented in three different configurations:

Table 4.1 Comparison of soft core processors

Category	Nios II (Fast Core)	MicroBlaze	Xtensa	OpenRISC 1200	LEON
Operating speed (MHz)	200	200	350	300	125
Architecture type	RISC (32 bit)	RISC (32 bit)	RISC (32 bit)	RISC (32 bit)	RISC (32 or 64 bit)
Custom instructions	Up to 256 Instructions	None	Unlimited	Unspecified limit	None
Pipeline stages	6	3	5	5	7
Register	32	32	32 or 64	32	2–32

- Nios II/f: f means "fast" version developed for best performance. This version uses more resources of FPGA but the system performance is much better.
- Nios II/s: s means "standard" version. This version of core is developed to maintain the equilibrium between the system performance and system cost. This also makes used of minimum chip recourses by sacrificing on better performance.
- Nios II/e: e means "economy" version. This processor version makes use of minimum amount of FPGA resources. This version has the less set of features which are user configurable and modifiable. For low-cost FPGA application, this version of core is more suitable.

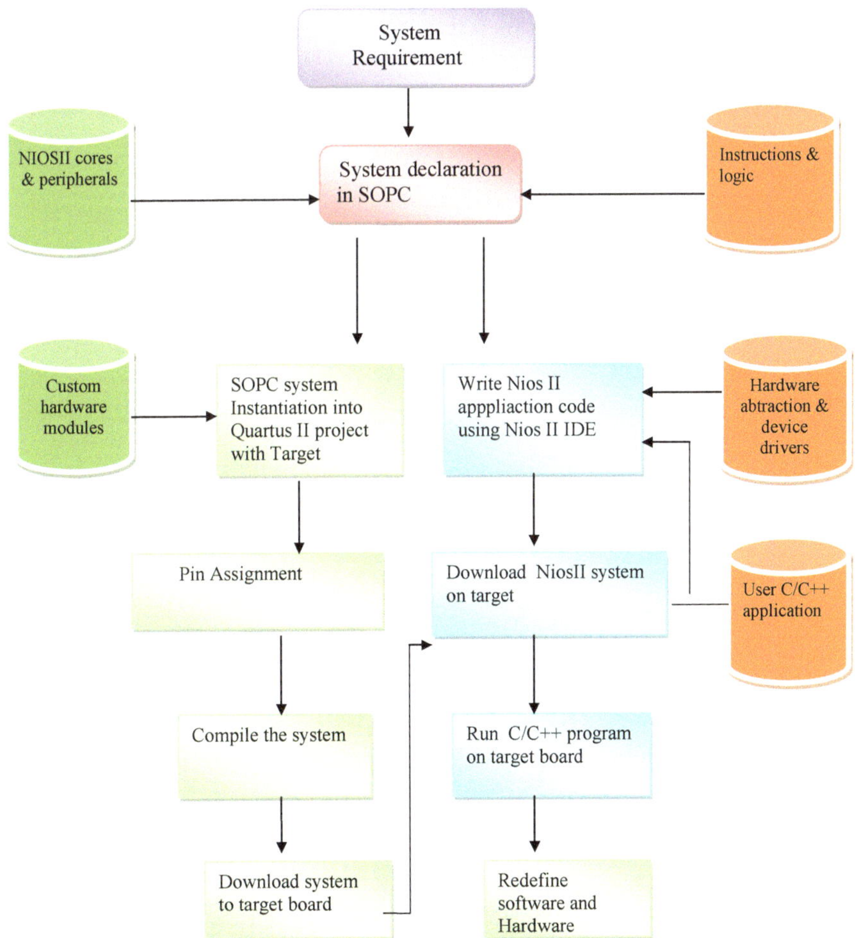

Fig. 4.1 Nios II system design development flow

The architecture of processor Nios II is based on RISC. Operands present in general-purpose registers are used to perform arithmetic and logic operations. Load/ Store instructions are used to move data between these general-purpose registers and memory.

Nios II is 32-bit processor have capability to work either in big-endian or little-endian mode.

Following are the three modes in which Nios II operates:

- **Supervisor mode**: At reset, processor enters supervisory mode that enables the processor to compile all instructions and execute all available functions.
- **User mode**: User mode is kind of restricted mode which prevents the handling of few instructions. In this mode, all the features of processor are not accessed.
- **Debug mode**: To implement features like watch points and breakpoints, one has to use software debug mode.

Designed application programs are possible to run in either the user or supervisor modes. The processor version currently available does not support the user mode.

4.5 Design Flow

Nios II System Design Development Flow

The design flow Nios II system development goes through three stages: system design, hardware design, and software design steps. For complex systems design application, two different designers team are required one handle hardware while other looks after software and also require proper coordination among themselves. Designs which are less complex in nature can be handled solely by single person.

The entire design and development flow for Nios II system creation and prototyping it is shown in Fig. 4.1.

Chapter 5
How to Build First Nios II System

Contents

Abstract This chapter gives an introduction to Altera's SOPC Builder, which is used for the implementation of system that uses the Nios II processor on an Altera FPGA device. The system development flow is illustrated by giving step-by-step instructions for using the System-On-a-Programmable-Chip (SOPC) Builder in conjunction with the Quartus II software (Version 7.2) to implement a desired system. The final step in the development process is to configure the circuit designed in a FPGA device and running a desired application program in C/C++ using Nios II IDE.

Keywords Nios II IDE · SOPC Builder · Configure

This chapter will make you familiar with Altera's tool SOPC Builder, which is used to build a Nios II processor system on an Altera FPGA device. The system design development flow is illustrated in detail by giving step-by-step instructions for using the SOPC Builder in conjunction with the Quartus II software (Version 7.2) to implement a desired system. The final step in the development process is to configure the circuit designed in a FPGA device and running a desired application program in C/C++ using Nios II IDE. To implement this, the user requires an access to a Altera DE2 Development board interface to a computer having Quartus II and

J.S. Parab et al., *Hands-on Experience with Altera FPGA Development Boards*,
https://doi.org/10.1007/978-81-322-3769-3_5

Nios II IDE software installed on it. Altera Nios II is a soft processor described in HDL, which will be implemented by using the Quartus II CAD system on Altera FPGA.

To generate desired system, it is compulsorily required to include functional units such as memory, parallel I/O ports, programmable timers, and communications interfaces. To implement such systems, it is useful to have CAD software to design a SOPC. Altera provides SOPC Builder as a part of Quartus II IDE. The following subsection will introduce a very user-friendly tool of Altera called SOPC Builder that allows a quick implementation of a simple Nios II system on DE2 development board of the Altera.

NIOS II System

One can implement a system using Nios II on embedded platform DE2 board as shown in Fig. 5.1.

The soft processor Nios II interacts with the other chips on the board (DE2) through the interfaces created in Cyclone II FPGA chip (i.e., the SRAM, SDRAM, and flash memory chips on the board DE2 are accessed through the appropriate interfaces created inside FPGA Cyclone II). All components shown above need to be connected by Avalon Switch Fabric interconnection network. Soft processor Nios II has memory which is on-chip made up of several Cyclone II memory blocks. Typical computer systems I/O ports are provided by Parallel and serial I/O interfaces. A special interface, i.e., JTAG UART is used to connect to the circuitry

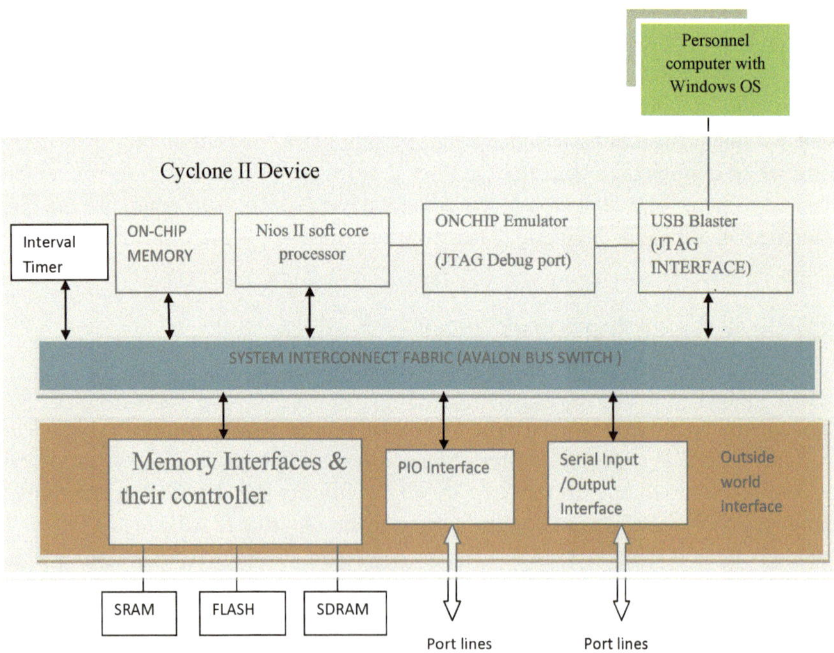

Fig. 5.1 A Nios II system implemented on the DE2 board

that provides a USB link to the host computer, which is called the *USB-Blaster*. JTAG Debug system module used here will control the Nios II system using host computer, which performs functions such as dumping programs, starting and ending execution, and fixing breakpoints. Using HDL language, components of Nios II system are defined and ported on FPGA. However, this would be a troublesome and highly time consuming. Instead, SOPC Builder can be used for desired system implementation, by selecting the component required. Here, we try to explore the SOPC Builder capability by designing very simple system. One can follow the same methodology to design large system.

Altera's SOPC Builder
Altera Quartus II CAD software is used in conjunction with SOPC Builder. This tool helps the designer to select the required modules to create a system using Nios II.

5.1 Creating the Advanced Quartus II Project

To implement the desired system, one has to start the Quartus II software and do the following simple steps:

Design system project is created in Quartus II as given in Fig. 5.2, (we stored our project in a location called *Testproject in D drive*, and we assigned project name as *lights,* and the same name is also assigned for the top-level design entity). One has to select a different project name, but one has to be very careful while assigning project/directory name since SOPC Builder software does not allow to file names with spaces (e.g., an attempt to use a directory name *Test Project* would lead to an error). In your project, EP2C35F672C6 chip is selected as the target device, since this is the FPGA present on the DE2 development board. However, one has the flexibility to choose other target device.

5.2 Creation and Generation of NIOS II System by Using SOPC Builder

Following steps will help you to get familiar with SOPC Builder and to create a simple Nios II system required to build Nios II system.

1. Choose Tools and then select option SOPC Builder, which will show you a box as shown in Fig. 5.3. Enter the system name; SOPC will generate a system with the same name. Choose VHDL in the selection, in which the system module will be specified. Click OK, then a window as shown in Fig. 5.4 appears. If you choose Verilog, then modules will be created in Verilog (We have selected VHDL).
   Tab with system contents is shown in Fig. 5.4 which helps for adding and configuring the selected components to meet the requirements of design.

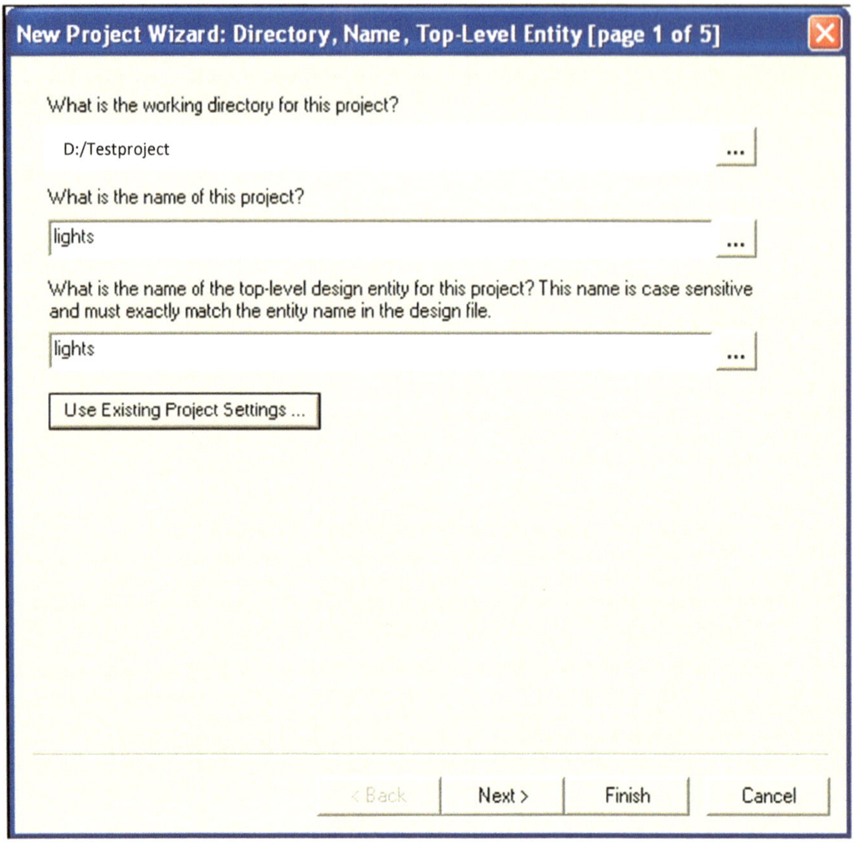

Fig. 5.2 Create a new project

Fig. 5.3 Creating a new Nios II system

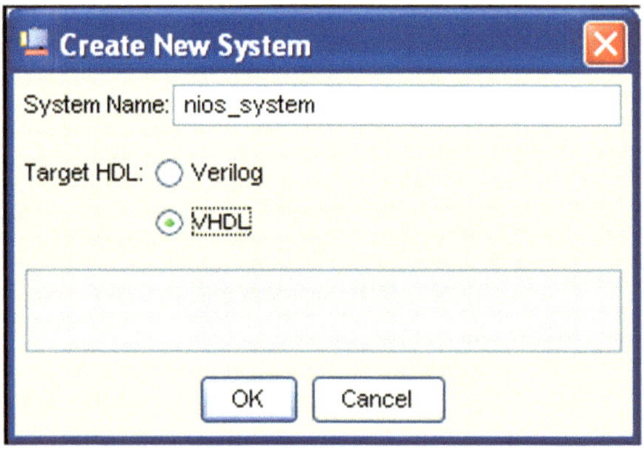

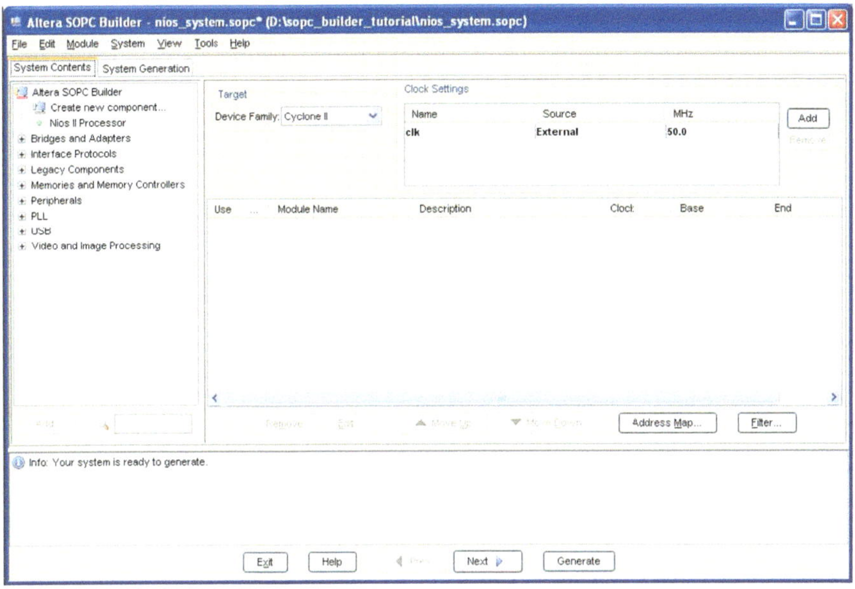

Fig. 5.4 Window showing system contents

Window on the left side shows available list of components. Before choosing any components, check the area in the figure-labeled target. Select the target available on the board as Cyclone II.

2. DE2 board is provided with 2 clock sources, but we are designing a Nios II system controlled by 50-MHz clock. As shown in Fig. 5.4, one has to specify a clock name as *clk* with designated source as external, and the frequency is set to 50.0 MHz.
3. Now, specify the processor as follows:
   - From the left-hand window pane, select **Nios II Processor** and click **Add**. Select **Nios II/s** for the processor core which gives window as displayed in Fig. 5.5.
   - Simplest and standard version of the processor Nios II/S is selected and then click Finish, which now displays the Nios II processor specified in Fig. 5.6. There will be a lot of warnings or error messages generated on console, since many parameters have not yet been specified. *Ignore these messages at this moment of time.*
4. Steps to include on-chip memory in system:
   - Choose Memories and Memory Controllers > On-Chip memory > On-Chip Memory and then press Add button.
   - In the Configuration Wizard window of On-Chip Memory, as shown in Fig. 5.7, set the width of memory as 32 bits.

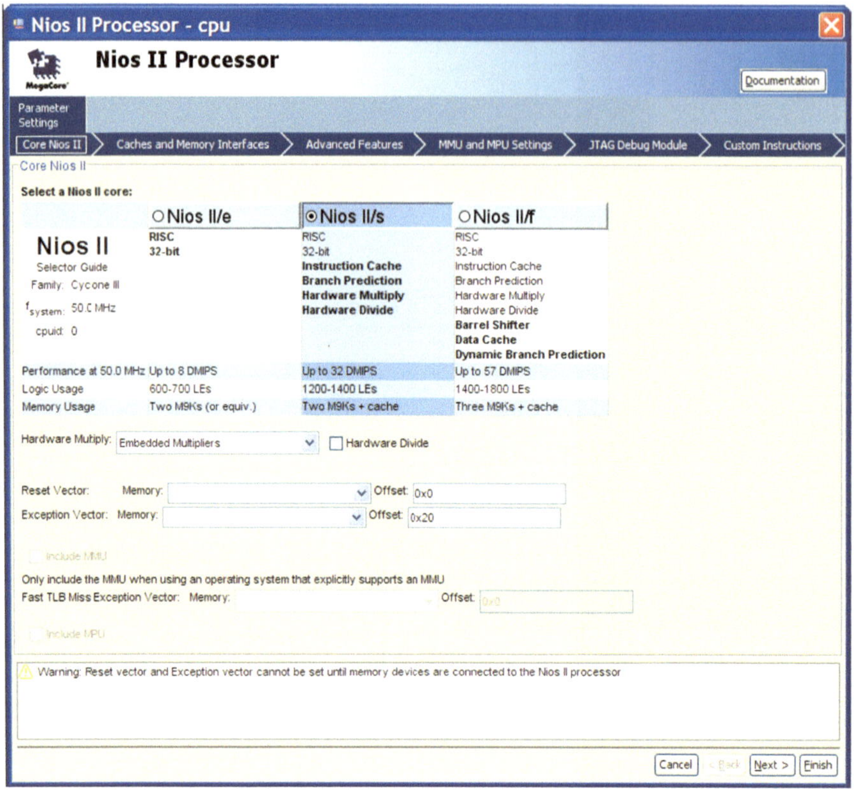

Fig. 5.5 Nios II processor selection

- Other default settings are unaltered.
- Then, click Finish that gives system contents tab as shown in Fig. 5.8.

5. Specifying PIO (Parallel Input/output) interface by following below steps:

   - Choose Peripherals > Microcontroller Peripherals > PIO (Parallel I/O) and click button Add to get the Configuration Wizard of PIO shown in Fig. 5.9.
   - Width and direction of the port is specified (8-bit port as input) as per user requirement as shown in Fig. 5.9.
   - Press finish to go back to system contents tab as shown in Fig. 5.10.

6. In the similar manner, assign the output I/O interface:

   - Choose Peripherals Tab > Microcontroller Peripherals > PIO and Press Add to get PIO settings.
   - Width and direction of the port is specified (8-bit port as output).
   - Press finish to complete the selection.

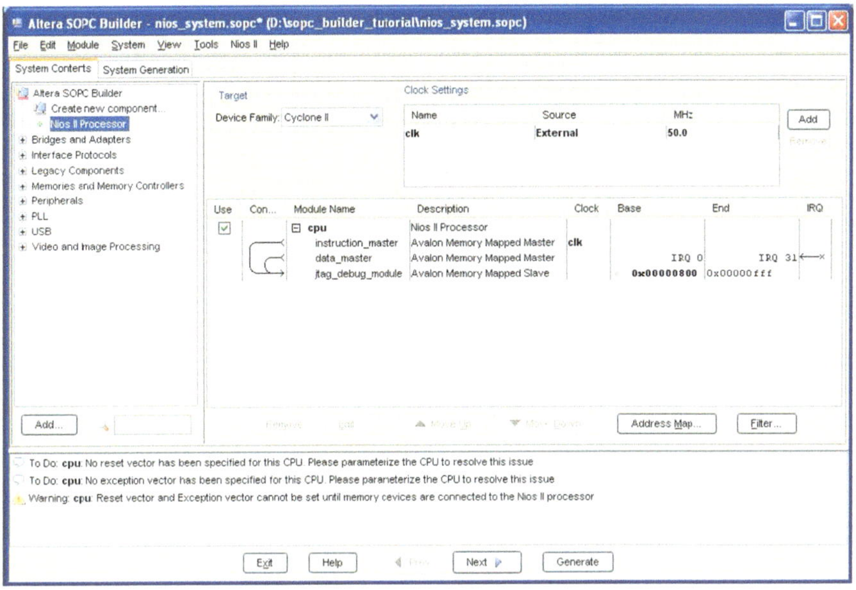

Fig. 5.6 NIOS II processor inclusion

7. To create an interaction between Designed Nios II system and host computer, some interface is required. This will be achieved by including the JTAG UART interface as given below:

   - Choose Interface Protocols > then select Serial > JTAG UART and press Add to get Configuration Wizard JTAG UART as shown in Fig. 5.11.
   - Do not alter the original default settings.
   - Press finish to complete the configuration.

8. The entire components of designed system are given in Fig. 5.12.
   SOPC is an intelligent tool of Quartus which automatically assigns names for the various components. Designer can change these names as an when required by Right clicking on the pio name and then select Rename.
9. SOPC Builder automatically assigned the base and end addresses for the components included in the designed system which can also be modified by the designer. To auto-assign base address click on > Auto-Assign Base Addresses, which gives the assignment as given in Fig. 5.13.
10. The settings of Nios II processor such as reset and exception vector addresses are specified by performing following:

    - Keep the mouse cursor on the cpu, then right click and then choose edit.
    - Choose onchip_mem as memory for both reset and exception vector, as depicted in Fig. 5.14.
    - Default offset setting is not altered.
    - Click Finish to complete the Nios II specification.

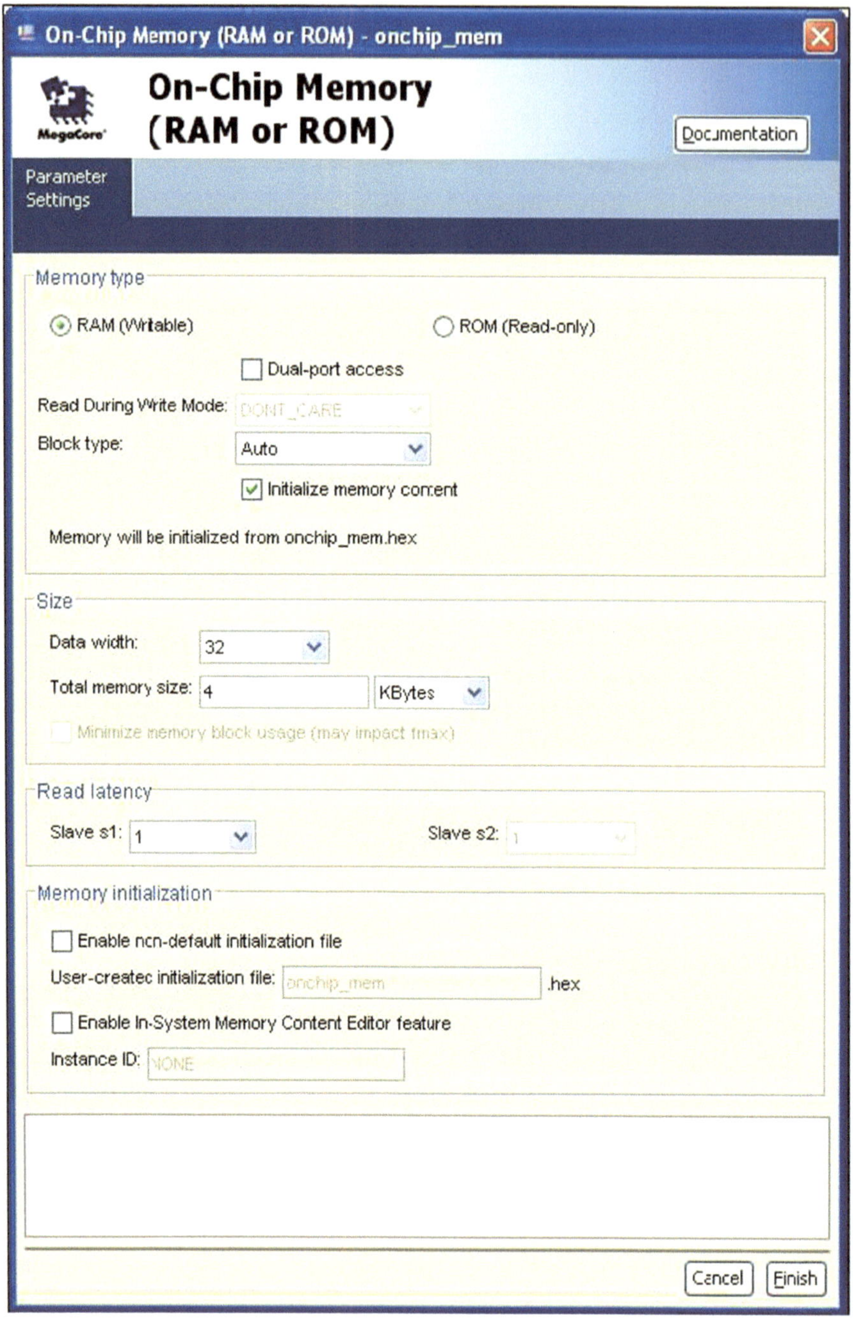

Fig. 5.7 Definition of the on-chip memory

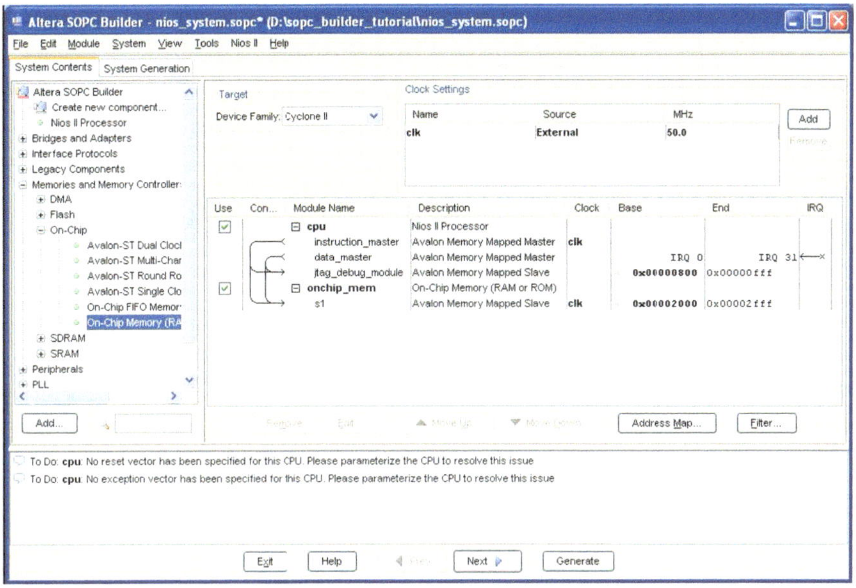

Fig. 5.8 System content tab with on-chip memory added

11. After specifying all components required to implement the desired system, the full system is now ready for generation. Choose the > System Generation tab, select Turn off Simulation—Create simulator project files. Click Generate on the bottom of the SOPC Builder window. After the generation process completion, messages as shown in Fig. 5.15 are displayed. Then, press Exit to returns to the main Quartus II window.

Any changes required to the designed system can be easily made at any point of time by opening SOPC Builder tool.

5.3 Nios II System Integration into a Quartus II Project

In the earlier chapter, we have seen VHDL/Verilog design entry method. Here, we choose schematic entry methods for the integration of the generated system module that depends on to the Quartus II project.

To complete the design, use the following steps:

- Instantiate the Nios II system module generated by the SOPC Builder into the Quartus II project by double clicking on Block Design File (BDF) file, then window as shown in Fig. 5.16 appears.

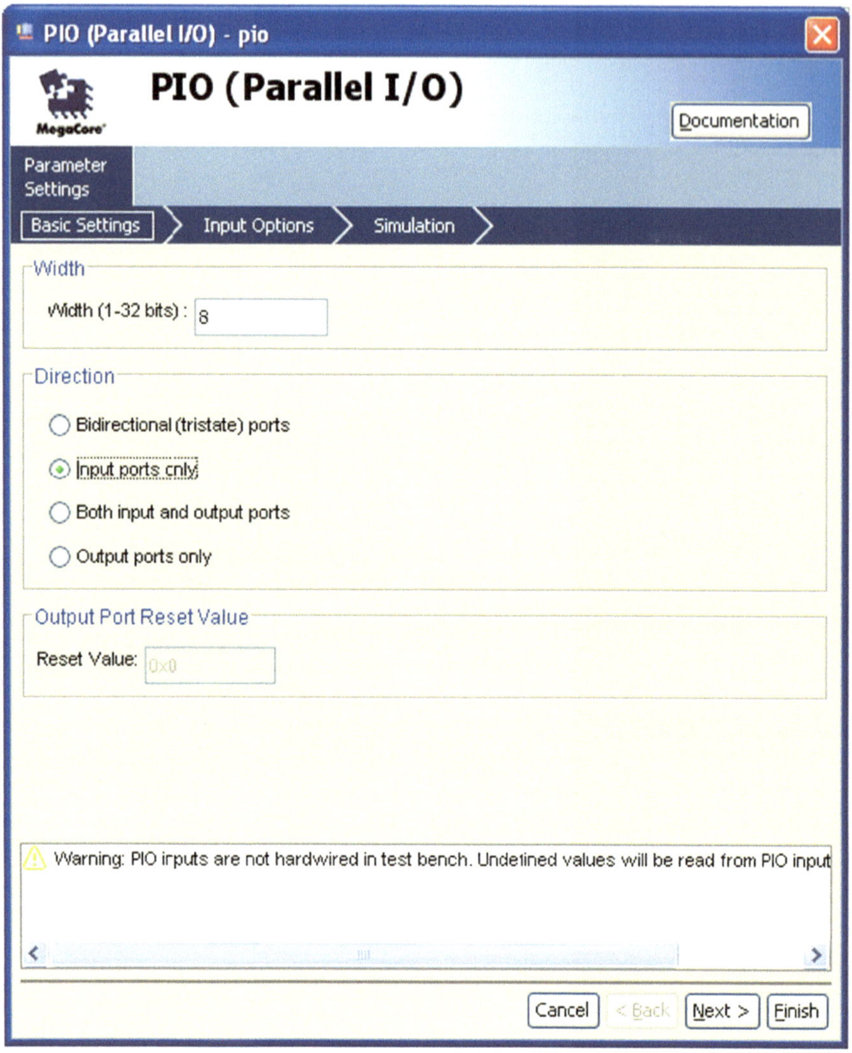

Fig. 5.9 Parallel input/output interface declaration

- One has to click Project and select the Nios II system by clicking ok and paste the system on BDF file.
- Now, next step is to generate the pins of Nios II system by right clicking on the system and select generate pins for symbol ports, the system will look like as shown below Fig. 5.17.
- Now, one has to rename the pins as provided in the DE2 pin assignment file (DE2_pin_assignments.csv) by double clicking on respective pin as shown in Fig. 5.18 and change the pin name, i.e., clk as CLOCK_50, reset_n as KEY[0],

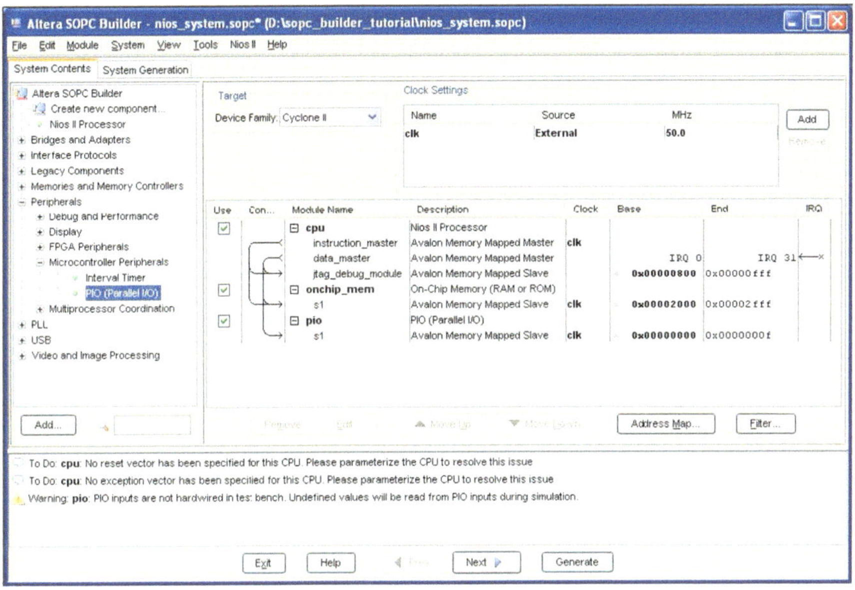

Fig. 5.10 Inclusion of PIO interface in system

in_port_to_the _switches[7…0] as SW[7…0] and out_port_from_the_LEDs as LEDG[7…].

- **Next Step is Assignment of FPGA Pins**:

There are two ways of assigning the pins manual pin assignment and automatic pin assignment:

When the system design is large and many input/output pins are involved, one has to select the automatic pin assignment model, whereas for simple design, one can go for manual pin assignment.

(1) Manual Pin Assignment:
Here, to see the pins in assignment editor directly, one has to compile the entire system by clicking the start compilation under the processing toolbar. Once the entire system is compiled without any errors (warnings generated are accepted), then go to Assignment → Pins (window appears as shown in below Fig. 5.19). To assign FPGA pin to Node (pin named in design file) click on the Location and select the respective pin of FPGA, continue this process till you assign all the pins.

(2) Automatic Pin Assignment:

- Go to Assignment → Import Assignment → Browse the DE2_pin_asignment.csv file and click on OK.

Fig. 5.11 JTAG UART module interface

- Next step is to generate the tcl script for the project by choosing Project > Generate Tcl file for project (Fig. 5.20) then click OK.
- To run the Tcl script select tools > script (Fig. 5.21) Then, click on Project Name and click on Run.

- **Compilation of Design**

Once the assignment is done, save the design file and select start compilation option by clicking on Processing tab → start compilation or by clicking on the toolbar icon ▶. As the compilation progresses through various stages, its progress is reported at left side of the Quartus II display.

After successful analysis and synthesis, compilation report is generated as shown in Fig. 5.22. If there is error, click on that error so that helps you for debugging the design.

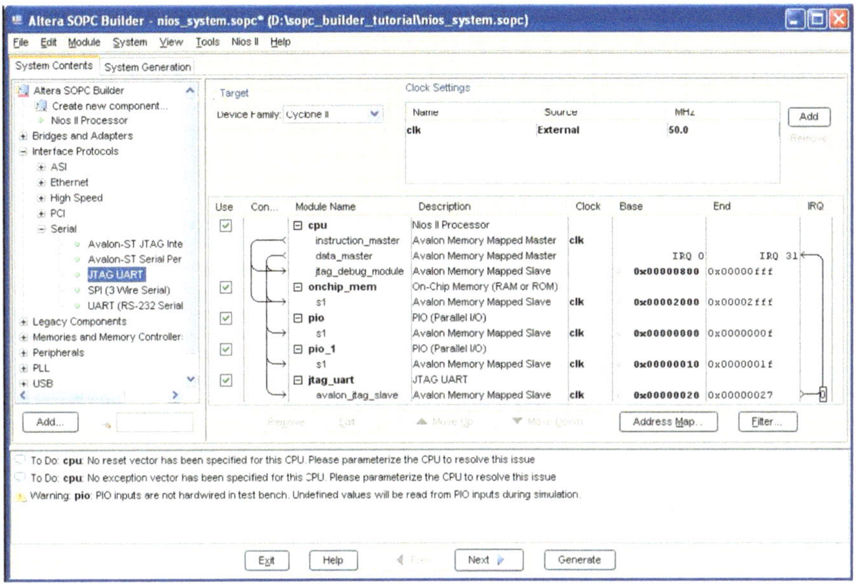

Fig. 5.12 Complete system

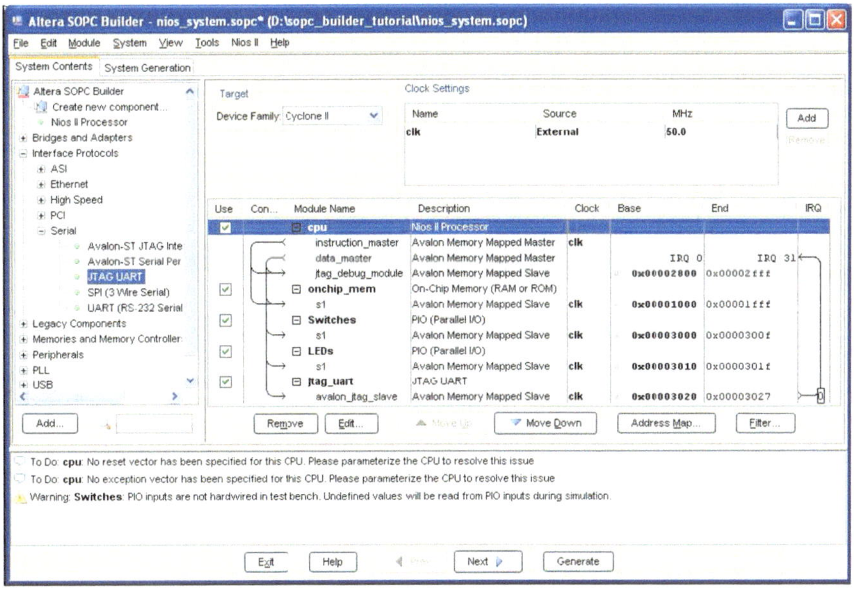

Fig. 5.13 Full proof final system specification

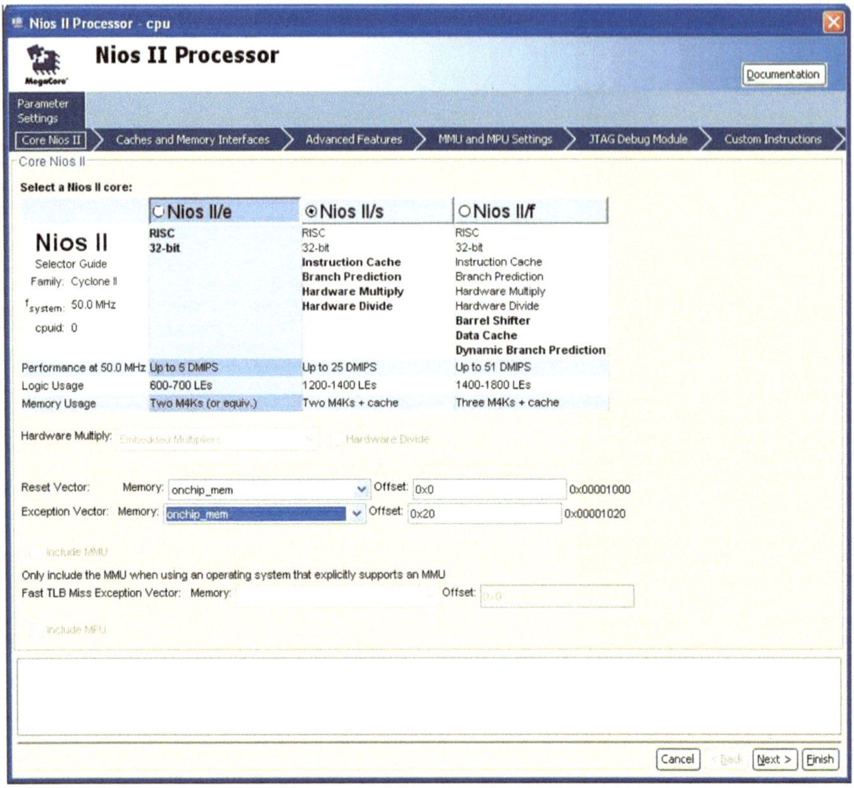

Fig. 5.14 Reset and exception vector declaration

When the compilation is finished, a compilation report is produced automatically. One can also open the compilation report, and it can be opened at any time either by selecting Processing > Compilation Report or by clicking on the icon .

The report includes a number of sections listed on the left side of its window.

5.4 Programming and Configuration Cyclone II Device on the DE2 Board

Program and configuring details of Cyclone II FPGA is explained in Chap. 2.

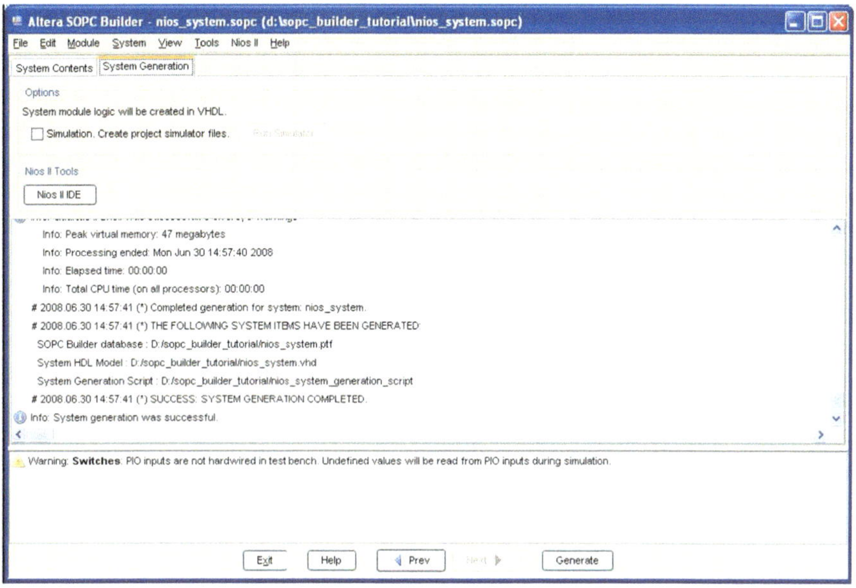

Fig. 5.15 Full generated system

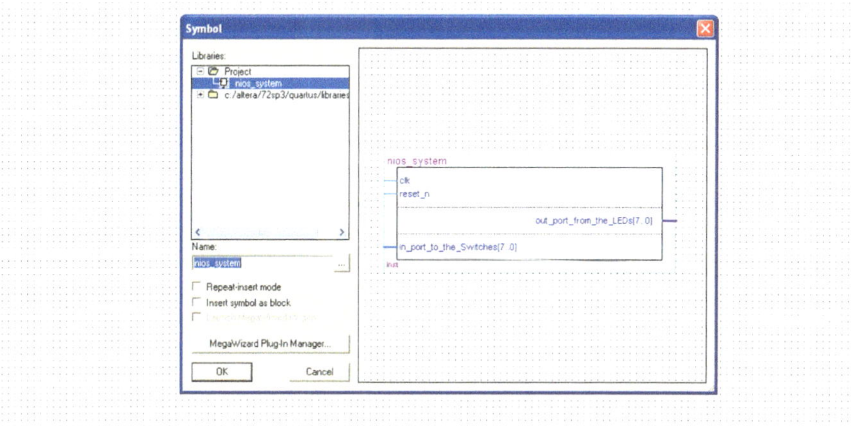

Fig. 5.16 Nios II system component

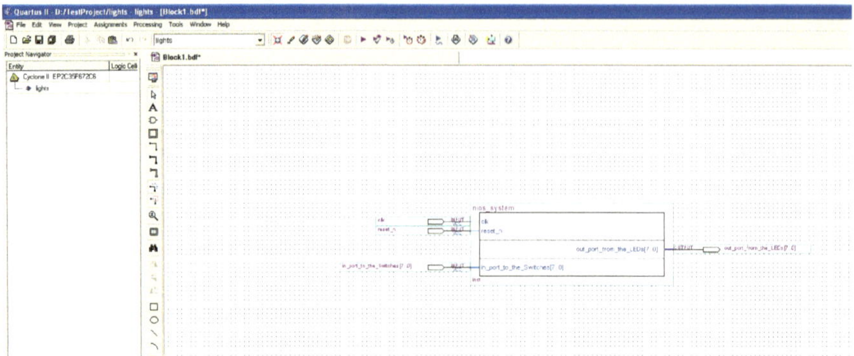

Fig. 5.17 Complete Nios II system in Quartus

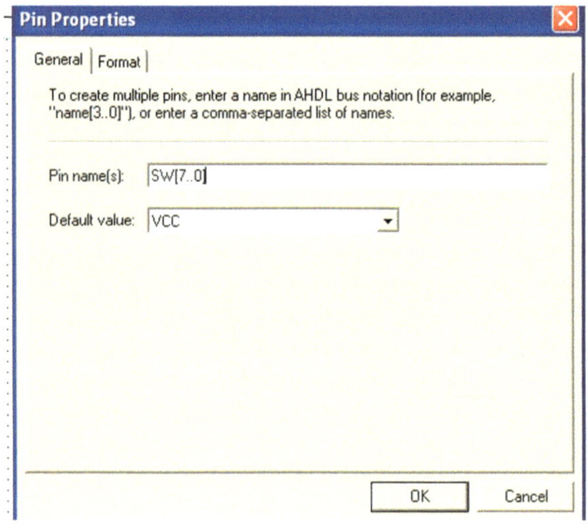

Fig. 5.18 Renaming the pins

5.5 Creating C/C++ Program Using Nios II IDE

5.5.1 *Introduction*

Nios® II integrated development environment (IDE) helps programmer to write his own C/C++ program which will control the different peripherals included in the designed system. This section will highlight some important features of Nios II IDE.

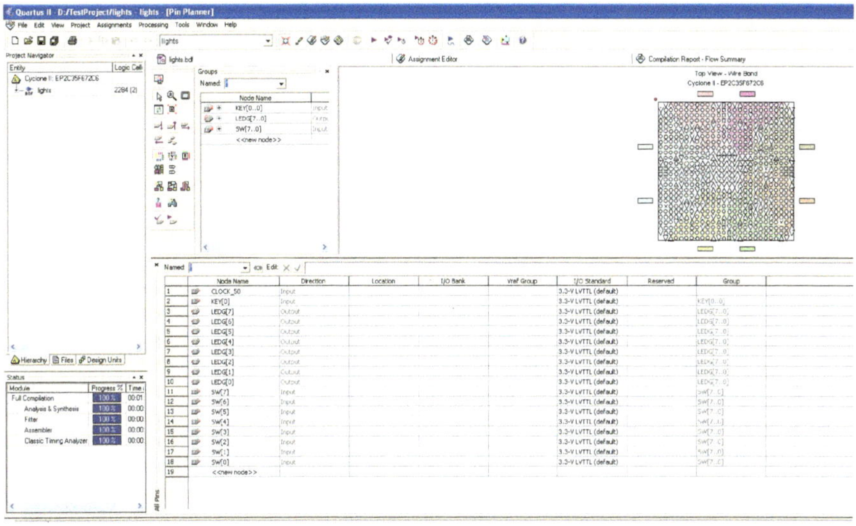

Fig. 5.19 Manual pin assignment window

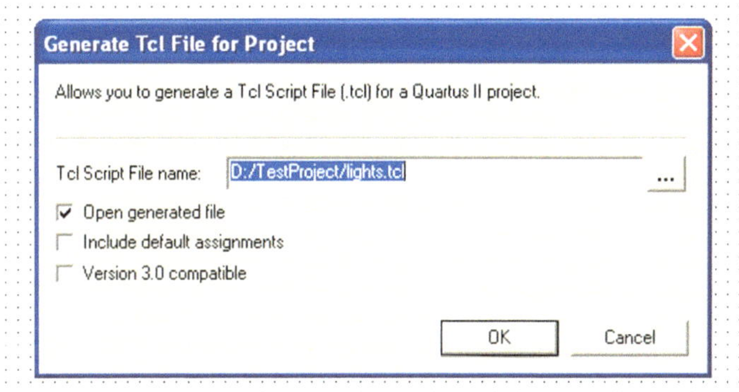

Fig. 5.20 Generating tcl script for project

The Nios II IDE Workbench

Nios II IDE workbench is a desktop development environment. The workbench is the place where you edit, compile, and debug your programs. The snapshot of how the Nios II IDE workbench looks is shown in Fig. 5.23.

Perspectives, Editors, and Views

Every perspective gives a set of capabilities for doing a specific task. Development perspective Nios II C/C++ IDE is depicted in Fig. 5.23.

Perspectives under workbench consist of editor slot and one or more views area. To open and edit a resource of project, an editor is used.

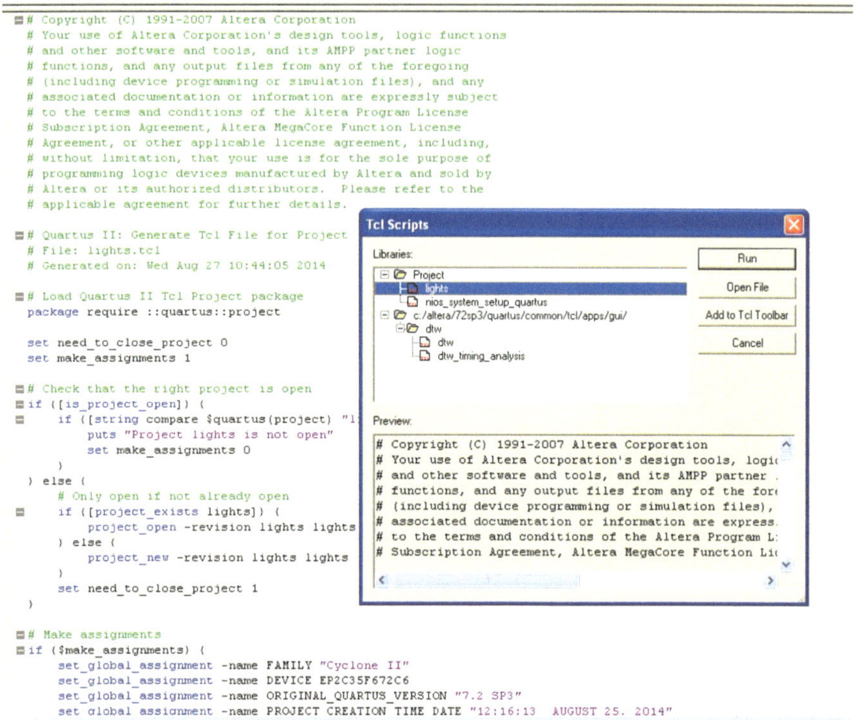

Fig. 5.21 Running the tcl script

Editor showing the C program and Project's view of Nios II C/C++ in the left-hand side of workbench is given in Fig. 5.23. View display of C/C++ Projects gives the content information about the active Nios II projects.

Programmer can open many editor windows, but at given time, only one can be active. Tabs in the editor area indicate the names of resources that are currently open for editing.

Creating a New IDE Project

Creating a Nios II IDE project is very simple; one has to follow the following steps carefully. Here, New Project wizard of IDE that guides you to create new IDE-managed projects. To start the New Project wizard, click on File menu, then hold cursor on New, and then choose Nios II C/C++ application as given in the below Fig. 5.24.

New Project wizard of Nios II IDE prompts to specify:

1. A name to new Nios II project.
2. The target CPU.
3. Project template.

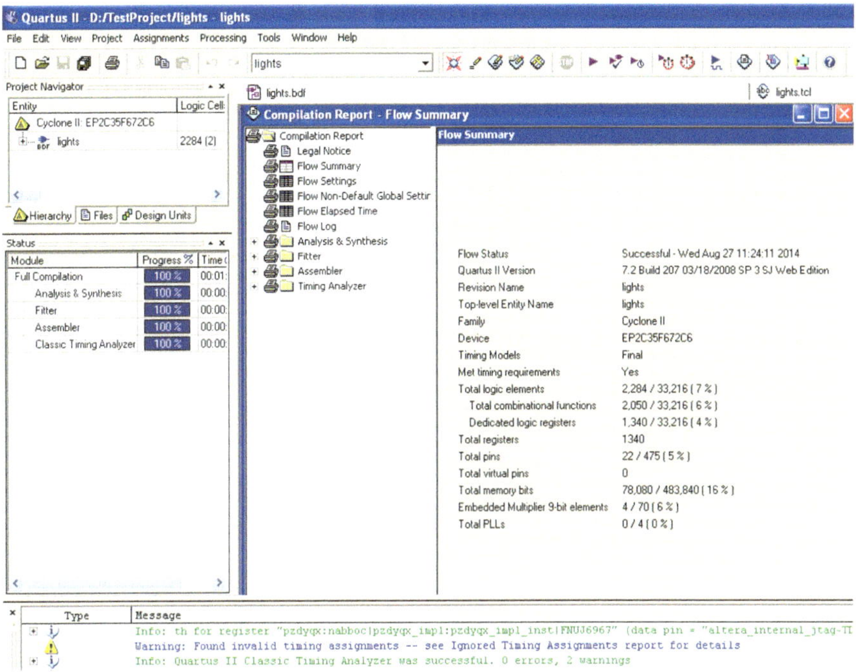

Fig. 5.22 Display after a successful compilation (compilation report)

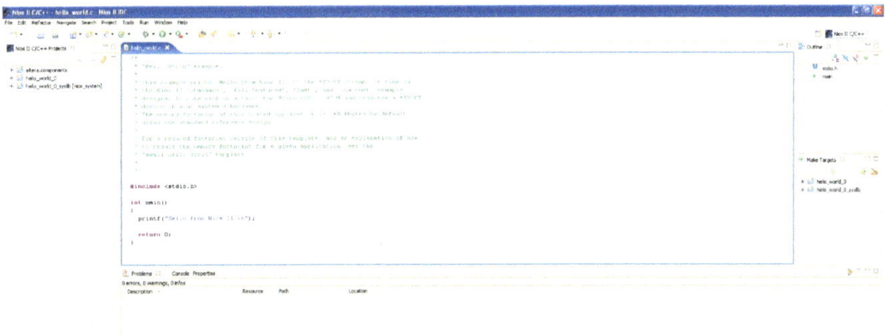

Fig. 5.23 Workbench space of Nios II IDE

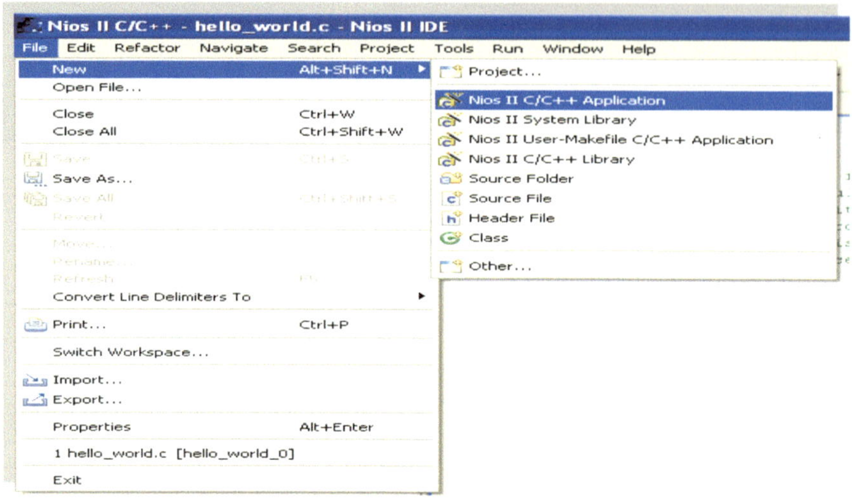

Fig. 5.24 New Project wizard Nios II C/C++ application

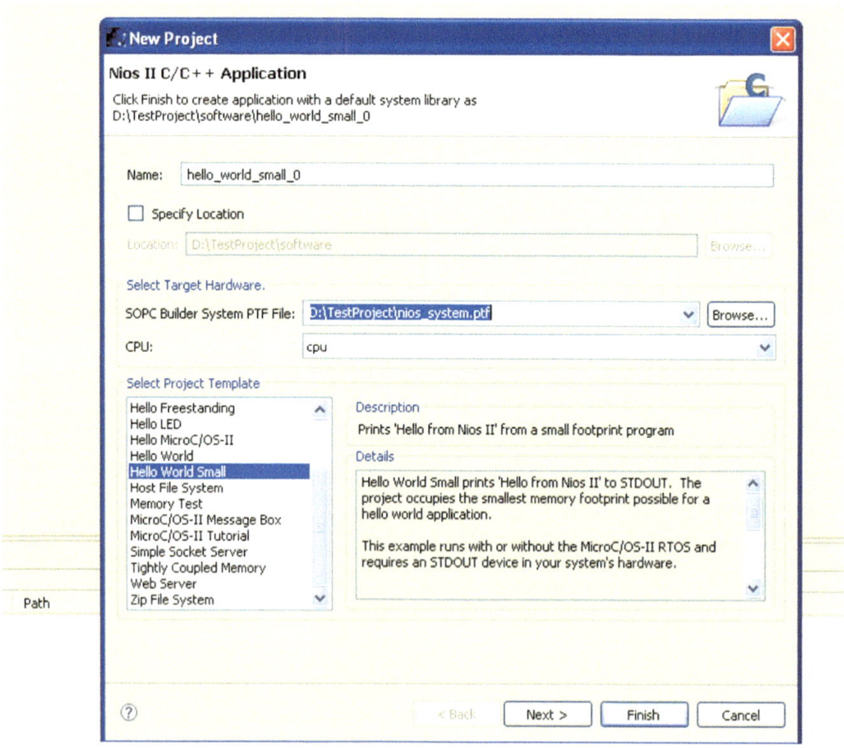

Fig. 5.25 Hello world template selection

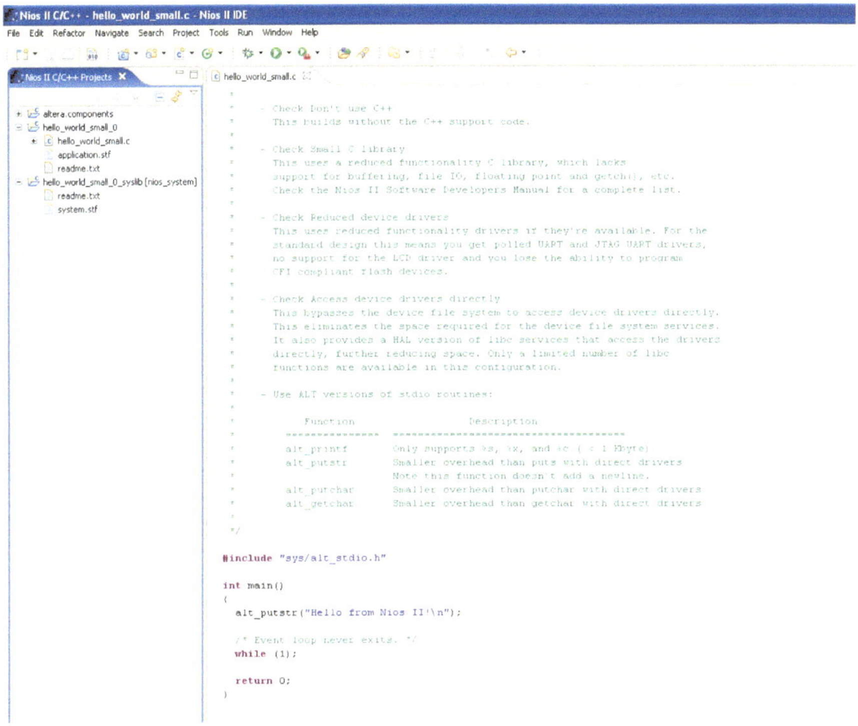

Fig. 5.26 Hello world Nios II IDE C, C++ project

It is always advisable to start with the Hello world small template. "Hello World small" template selection is shown in Fig. 5.25.

Then, click Finish which creates the new project (Fig. 5.26); it also generates system library for the project.

5.6 Running and Testing It on Target Board

Building Projects

Although the commands are available on tool bar menu, but right clicking is the fastest way to locate the required commands.

To compile a project, hold the cursor on the project, right click on the project, and press Build Project. Figure 5.27 shows how to get the build project option. First, system library project is generated, and then, entire project is compiled.

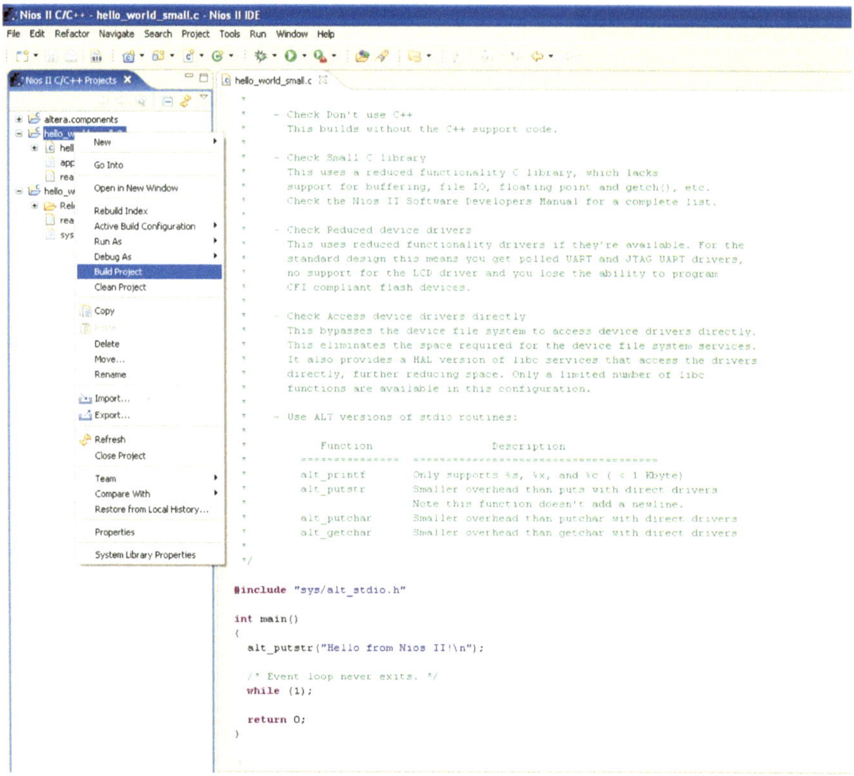

Fig. 5.27 Right click on Project to get build option

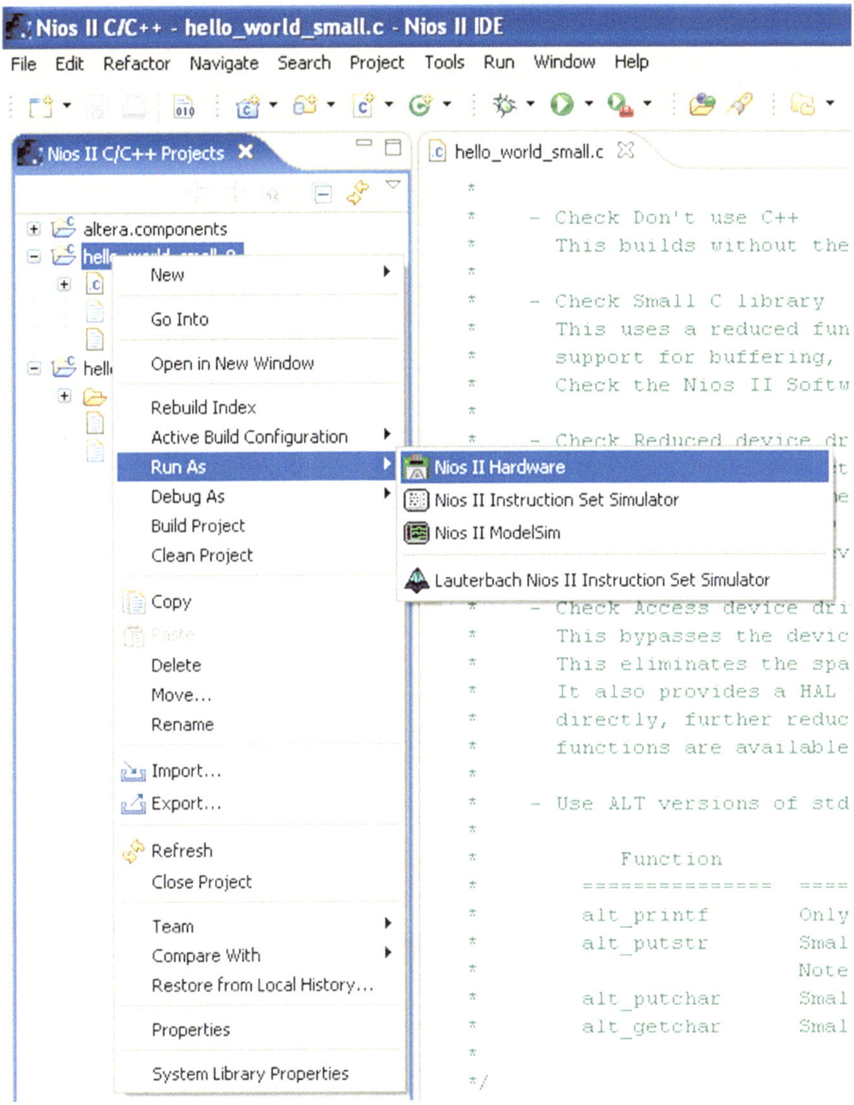

Fig. 5.28 Running a program on target hardware

Running and Debugging Programs

User can run or debug the project on target board or NIOS II instruction simulator (Fig. 5.28). To run the code on board, one has to right click on the Project, then choose Run As > then select NIOS II Hardware. This runs the code on target boards and displays the desired result on the board.

If the entire system and written code is correct, it will display a message on console as **"Hello from Nios II"**.

Chapter 6
Case Studies Using Altera Nios II

Contents

Abstract This chapter will further boost the interest as it covers lots of interesting case studies designed around Nios II soft core processor such as blinking of LEDs in different patterns, displaying scrolling text on LCD, interfacing camera for acquiring images, multiprocessor communication, Ethernet-based robotic arm control, matrix crunching problem for multivariate analysis, and reading flash for Web application.

Keywords Multivariate analysis · Robot control · Web application Multiprocessor communication · Camera interfacing

A Nios II-based embedded system design consists of customized hardware and software. To configure the processor and I/O peripherals, Altera's SOPC Builder tool is used and Nios II EDS platform is used to design software which runs on the designed hardware. We have already explained in Chap. 5 detailed procedure for creating an Nios II system. In this chapter, we will provide in brief how to create an Nios II system for particular application on how the hardware and software interface and basic coding techniques help to access low-level I/O peripherals.

J.S. Parab et al., *Hands-on Experience with Altera FPGA Development Boards*,
https://doi.org/10.1007/978-81-322-3769-3_6

6.1 Blinking of LEDs in Different Patterns

Light-emitting diodes are the commonly used components in many applications to display the different sequences. The DE2 board which we are using has 26 LEDs which are user-controllable: 18 are red LEDs, and 8 are green LEDs. Every LED is driven by Cyclone II FPGA pin directly; sending high logic level to pin turns the LED on, and driving low on the pin turns it off. Here, we have selected eight green LEDs to display different patterns.

To demonstrate the entire process, we have designed a simple blinking LED system run on soft core processor Nios II platform. The key steps in brief for the development of entire system for blinking of LEDs are as follows:

- Open the Quartus software and create a new project.
- Go to assignments select import assignments and add the de2_pin assignment file.
- Select create tcl file for project from the project menu and the run the tcl script by selecting tcl script from the tool menu.
- To select the components, open the SOPC Builder and choose the following components. After selection of below components, the complete SOPC Nios II system looks like as shown in Fig. 6.1.
    - NIOS II PROCESSOR (STANDARD)
    - JTAG UART
    - SRAM (512 KB)
    - PIO (RENAME AS led)
- Next auto-assign base addresses and irq.
- In the system generation tab, click generate.

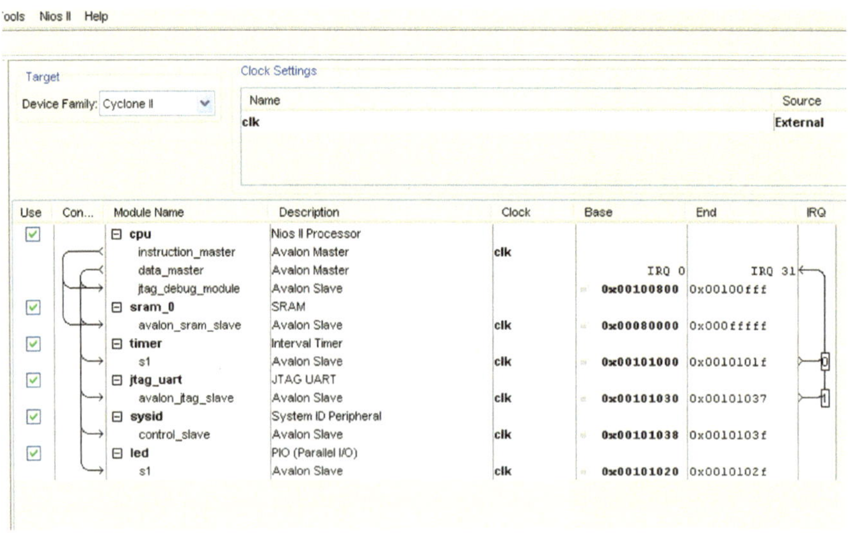

Fig. 6.1 Complete SOPC system

- Move back to the Quartus software and go to file menu and choose new block and schematic file.
- Right click on the workspace and add the component created in SOPC (located in the project folder).
- Add the respective connectors to the I/O generated in the component, and then, the entire system looks like as shown in Fig. 6.2.
- Save the entire project with the same file name as the entity and compile, and the compilation report is given in Fig. 6.3.
- In tools, choose programmer, check the program, configure tab, and click start.
- Open the Nios II IDE software.
- In the file menu, choose new C/C++ program, and from the template, choose blank project.
- Type in the following code mentioned below (led.c).

```
------------   blinking of LEDs C code ---------------------

#include <stdio.h>
#include <unistd.h>
#include "system.h"
#include "altera_avalon_pio_regs.h"
#include "alt_types.h"

alt_u8 led1[8] ={0x01,0x02,0x04,0x08,0x10,0X20,0x40,0x80} ;
int i,j;

int main()

{
  printf("THIS IS LED BLINKING PROGRAM\n");
for(i=0;i<6;i++)
      {
IOWR_ALTERA_AVALON_PIO_DATA(LED1_BASE, 0x00);
    usleep(1000000);
    IOWR_ALTERA_AVALON_PIO_DATA(LED1_BASE, 0xFF);
    usleep(1000000);
      }

while (1)
  {
      for(i=0;i<8;i++)
      {
IOWR_ALTERA_AVALON_PIO_DATA(LED1_BASE, led[i]);
   usleep(1000000);
      }
   for(i=7;i>=0;i--)
      {
IOWR_ALTERA_AVALON_PIO_DATA(LED1_BASE, led[i]);
    usleep(1000000);

 }
   }
   }
```

- Build the entire program and run on hardware to see the LED blinking effect.

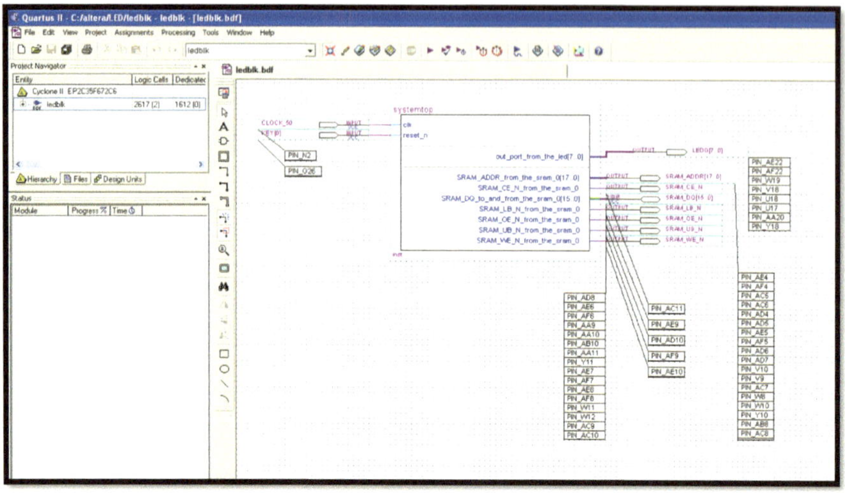

Fig. 6.2 Complete Nios II system in Quartus

Flow Status	Successful - Wed Apr 22 11:18:53 2015
Quartus II Version	7.2 Build 207 03/18/2008 SP 3 SJ Web Edition
Revision Name	ledblk
Top-level Entity Name	ledblk
Family	Cyclone II
Device	EP2C35F672C6
Timing Models	Final
Met timing requirements	Yes
Total logic elements	2,623 / 33,216 (8 %)
Total combinational functions	2,357 / 33,216 (7 %)
Dedicated logic registers	1,612 / 33,216 (5 %)
Total registers	1612
Total pins	53 / 475 (11 %)
Total virtual pins	0
Total memory bits	46,208 / 483,840 (10 %)
Embedded Multiplier 9-bit elements	4 / 70 (6 %)
Total PLLs	0 / 4 (0 %)

Fig. 6.3 Compilation report

6.2 Display of Scrolling Text on LCD

LCD (liquid crystal display) is the modern technology used for various displays in mobiles, notebook, and Tablets PC. Like LEDs and plasma devices, LCDs allow displays to be much thinner than cathode ray tube (CRT) technology. One of the greatest advantages of LCDs is that it consumes very low power than LED and gas-display displays because they work on the principle of blocking light rather than emitting it. LCD displays consist of two plates of polarizing material with a special kind of liquid between them which has very high impedance. An electric current passed through the liquid causes the crystals to align, and it turns opaque so that light cannot pass through them.

Optrex LCD Controller 16207 Core is present on DE2 board with Avalon® interface which provides the hardware component interface and driver required to display characters using Nios® II processor on LCD panel. Device drivers are provided in the HAL system library for the Nios II processor. SOPC Builder has readily available LCD controller, and it can be easily integrated in SOPC Builder-generated system.

Functional Description

The LCD hardware comprises of 11 signals that connect to the pins of Optrex 16207 LCD panel—these signals are defined in the data sheet of Optrex 16207.

- Enable (output)—E
- Register Select (output)—RS
- Read or Write (output)—R/W
- Data Bus (bidirectional)—DB0–DB7

Figure 6.4 shows LCD controller core interface diagram.

Instantiating the Core in SOPC Builder

To select the components, open the SOPC Builder and choose the following components. After selection of below components, the complete SOPC Nios II system looks like as shown in Fig. 6.5.

- NIOS II PROCESSOR (STANDARD)
- JTAG UART
- SRAM (512 KB)
- LCD (16 × 2)

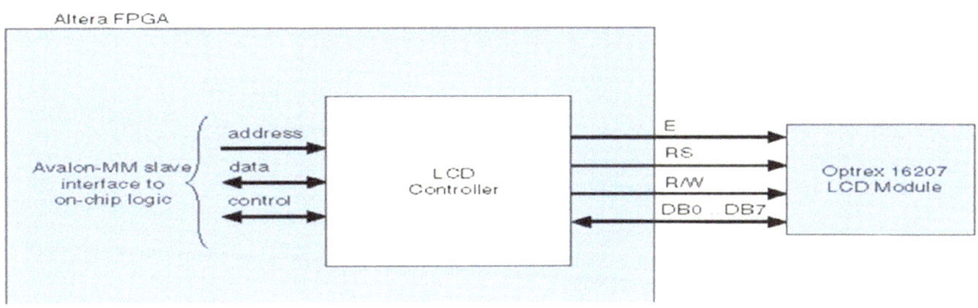

Fig. 6.4 LCD controller interface block diagram

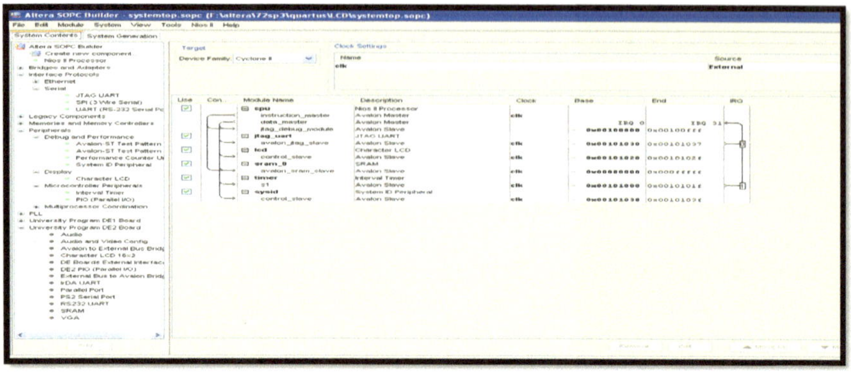

Fig. 6.5 SOPC components

In SOPC Builder, the LCD controller component has the name Character LCD (16 × 2, Optrex 16207). There are no user-configurable settings for LCD controller. The only choice one has to make in SOPC Builder is whether to add an LCD controller to the system or not. For each LCD controller added in the system, the top-level system module has 11 signals that connect to the LCD module.

Full-Fledged Nios II System for Scrolling LCD Display

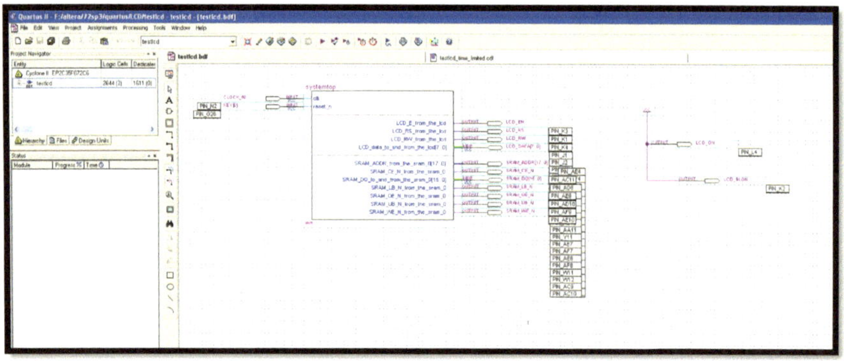

Compilation Report

Flow Status	Successful - Thu Jan 08 12:35:51 2015
Quartus II Version	7.2 Build 207 03/18/2008 SP 3 SJ Web Edition
Revision Name	testlcd
Top-level Entity Name	testlcd
Family	Cyclone II
Device	EP2C35F672C6
Timing Models	Final
Met timing requirements	Yes
Total logic elements	2,644 / 33,216 (8 %)
Total combinational functions	2,387 / 33,216 (7 %)
Dedicated logic registers	1,611 / 33,216 (5 %)
Total registers	1611
Total pins	58 / 475 (12 %)
Total virtual pins	0
Total memory bits	46,208 / 483,840 (10 %)
Embedded Multiplier 9-bit elements	4 / 70 (6 %)
Total PLLs	0 / 4 (0 %)

----------- C Code for scrolling tect on LCD display-------------

```
void lcd_init();
void LCD_show_Text(char*Text);
void LCD_Line2();
void LCD_Test();
//-------------------------------------------------------------------
//#endif
 #include<stdio.h>
 #include<unistd.h>
 #include"system.h"
 #include"alt_types.h"
 #include"altera_avalon_lcd_16207_regs.h"


 void lcd_init()
{
 usleep(15000);
IOWR_ALTERA_AVALON_LCD_16207_COMMAND(LCD_BASE,0X38);
usleep(4150);
IOWR_ALTERA_AVALON_LCD_16207_COMMAND(LCD_BASE,0X06);
usleep(4150);
IOWR_ALTERA_AVALON_LCD_16207_COMMAND(LCD_BASE,0X0E);
usleep(4150);
IOWR_ALTERA_AVALON_LCD_16207_COMMAND(LCD_BASE,0X01);
usleep(2050);
}
void LCD_Show_Text(char* Text)
{
  int i;
  for(i=0;i<25;i++)
  {
    IOWR_ALTERA_AVALON_LCD_16207_DATA(LCD_BASE,Text[i]);
    usleep(200000);
   if(i==15)
   {

  lcd_init();
    }
  }

  for(i=16;i<40;i++)
 {

  IOWR_ALTERA_AVALON_LCD_16207_DATA(LCD_BASE,Text[i]);
  usleep(200000);

}}
```

```
//----------------------------------------------------------------------------
void LCD_Line2()
{
  IOWR_ALTERA_AVALON_LCD_16207_COMMAND(LCD_BASE,0xC0);
  usleep(2000);
}
//----------------------------------------------------------------------------
void LCD_Test()
{
  char Text1[24] = "<This is Goa University>";

  lcd_init();

  LCD_Show_Text(Text1);

}
//----------------------------------------------------------------------------
int main (void)

 {
    printf ("abcd hello...... \n");

    while(1)
    {
        LCD_Test();
        lcd_init();

    }
 }
```

6.3 Interfacing of Digital Camera

Digital camera takes photographs by recording images electronically via an image sensor, and the digitized image is stored in a flash memory card. The image sensor used contains millions of pixel which helps in generating the proper image. A pixel sensor converts light to an electronic signal. The output of the pixel sensors is digitized and stored as an image file. A typical digital camera contains a set of buttons and knobs to control and adjust camera operation and a small LCD display to preview the stored pictures. The embedded system in the camera performs two major tasks. The first task involves the general "housekeeping" I/O operations, including processing the button and knob activities, generating the graphic on an LCD display, and writing image files to the storage device.

Connect D5M to Your DE2 Board as Shown in Below Figure

Main Features of D5M Camera Module

- High frame rate
- Low dark current
- Programmable control for gain, frame rate, frame size
- Automatic black level calibration

Key Performance Parameters

Parameter		Value
Active pixels		2,592 H × 1,944 V
Pixel size		2.2 μm × 2.2 μm
Color filter array		RGB Bayer pattern
Shutter type		Global reset release (GRR)
Maximum data rate/master		96 Mp/s at 96 MHz
Frame rate	Full resolution	Programmable up to 15 fps
	VGA (640 × 480)	Programmable up to 70 fps
ADC resolution		12-bit
Responsivity		1.4 V/lux-s (550 nm)
Pixel dynamic range		70.1 dB
SNRMAX		38.1 dB
Supply voltage	Power	3.3 V
	I/O	1.7–3.1 V

Connection Setup of DE2 Board

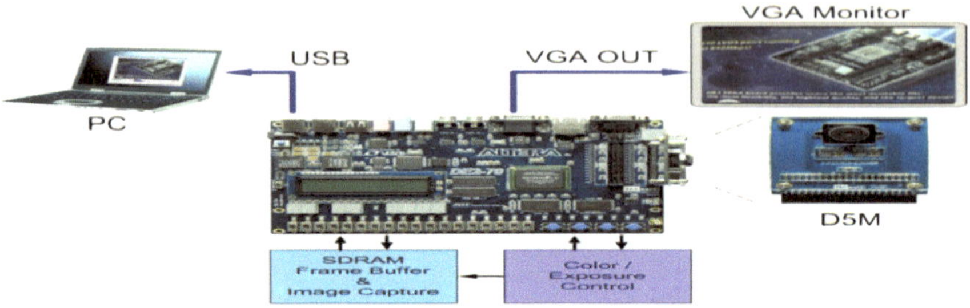

Procedure for Configuring the Digital Camera

When users buy the camera module from Terasic technologies, they provide CD containing the user manual and demonstration design files.

- Ensure the connection is done as shown in above figure. Make sure D5 M module is connected to GPIO1.
- Create project in Quartus and then click on SOPC to select the SOPC components as shown below
    - NIOS II
    - Avalon Trisatate Bridge
    - Audio and Video config under University Program DE2 Board
    - Interval Timer
    - JTAG Uart
    - RS 232 UART
    - SRAM
    - VGA
    - PIO (Input-8, Output-8).
- Then, generate the system and paste the system in Quartus Bdf file which looks like as shown in below figure.

- Compile the entire design and download the .sof file on to the DE2 board.
- Then, run the C code given below on Nios II system.

```
--------C Code to configure the came an capture the image and save the Image--------
NOTE: If Readers want the header file used in program they can contact the authors

#include "my_includes.h"
#include "camera_hal.h"
#include "packet.h"
#include "function.h"
#include "jtaguart.h"
#include "uart.h"
#include "LCD.h"



#ifdef DEBUG_APP
    #define APP_DEBUG(x)     DEBUG(x)
#else
    #define APP_DEBUG(x)
#endif
```

```
void init(void){

    UART_Open();

    APP_DEBUG(("\r\n===== DE2-70 Camera Utility [10/26/2007]
=====\r\n"));
    APP_DEBUG(("sizeof(PKLEN_TYPE)=%d\r\n", sizeof(PKLEN_TYPE)));

    if (!JTAGUART_Open()){
        APP_DEBUG(("JTAG open fail\r\n"));
    }else{
        APP_DEBUG(("JTAG open success and clear input buffer\r\n"));
        JTAGUART_ClearInput();
    }

    if (LCD_Open()){
        LCD_TextOut("Welcome DE2-70\nCamera Demo\n");
    }else{
        APP_DEBUG(("[LCD] open fail\r\n"));
    }

}

int main()
{
    alt_u32 time_start, ticks_per_sec;
    alt_u8 *szPacket;

    init();

    szPacket = malloc(PKT_NIOS2PC_MAX_LEN);
    if (!szPacket){
        APP_DEBUG(("malloc fail, program is terminated!\r\n"));
        return 0;
    }else{
        APP_DEBUG(("malloc %d byte success\r\n",PKT_NIOS2PC_MAX_LEN));
    }

    ticks_per_sec = alt_ticks_per_second();
    while(1){
            if (!read_packet(szPacket))
                continue;
```

```
        bool bResponse = TRUE, bSuccess = FALSE;
        time_start = alt_nticks();
        alt_u8 OP = szPacket[PKT_OP_INDEX];
        dump_op_name(OP);
        switch(OP){
            case OP_POLLING:
                bSuccess = op_polling(szPacket); // ack
                break;
            case OP_CAMERA_CONFIG:
                bSuccess = op_camera_config(szPacket);
                break;
            case OP_CAMERA_CAPTURE:
                bSuccess = op_camera_capture(szPacket);
                break;
            case OP_CAMERA_PORT_READ:
               bSuccess = op_camera_port_read(szPacket);
                break;
            case OP_MEMORY_READ:
                bSuccess = op_memory_read(szPacket);
                break;
            case OP_MEMORY_WRITE:
                bSuccess = op_memory_write(szPacket);
                break;
            default:
                bResponse = FALSE;
                break;

            }
            if (bResponse){
                alt_u32 time_elapsed;
                PKLEN_TYPE pl_len;
                alt_u32 pk_len;
            memcpy(&pl_len, &szPacket[PKT_LEN_INDEX], sizeof(pl_len));
// payload len
                pk_len = pl_len + PKT_NONEPL_SIZE;
                if (pk_len > PKT_NIOS2PC_MAX_LEN){
                    APP_DEBUG(("response packet len too long\r\n"));
                }else{
                    //DEBUG_PRINTF("pk_len=%d", pk_len);
                    //DEBUG_HEX_PRINTF(szPacket, pk_len);
                 APP_DEBUG(("JTAGUART_Write (len=%d)...\r\n", pk_len));
                    if (!JTAGUART_Write(szPacket, pk_len)){
                   APP_DEBUG(("send packet fail, len=%d\r\n", pk_len));
                    }
                }
                time_elapsed = alt_nticks() - time_start;
                APP_DEBUG(("\r\n%s(OP=%d, %d ms)\r\n",
bSuccess?"ok":"ng", OP,  (int)(1000*time_elapsed/ticks_per_sec)));
    return 0;

}
```

- Connect the output of DE2 board, i.e., VGA to VGA compatible monitor.
- Press KEY3 to make camera in FREE RUN mode.
- Press KEY2 to take a shot of photograph and then press KEY3.
- Below table summarizes the functions of keys.

Component	Function description
KEY[0]	Reset circuit
KEY[1]	Set the new exposure time (use with SW[0])
KEY[2]	Trigger the image capture (take a shot)
KEY[3]	Switch to free run mode
SW[0]	Off: Extend the exposure time On: Shorten the exposure time
SW[16]	On: ZOOM in Off: Normal display
HEX[7:0]	Frame counter (display ONLY)

6.4 Multiprocessor Communication for Parallel Processing

Multiprocessor system is a system which incorporates two or more processors working together to do one or more related tasks. Multiprocessor systems that share resources can be easily developed by using the Altera Nios II processor under SOPC Builder tool.

Multiprocessor Systems Benefits

The benefit of multiprocessor systems is increased performance at the price of significantly increased complexity of system. For such reason, multiprocessor systems have been limited to workstation and high-end PC computing use by using a symmetric multiprocessing (SMP) method of load-sharing. Usually, the overhead of SMP is too high for many embedded systems, but embedded platform-based application comprising multiple processors is gaining popularity since it performs different tasks and functions on different processors. Ideal platform for embedded multiprocessor systems can be developed using Altera FPGAs. Altera FPGAs make it possible to design system with many Nios II processors on a single chip. Here, different configurations of system can be designed, built, and evaluated very quickly by using SOPC Builder tool.

SOPC Components to Create Multiprocessor System

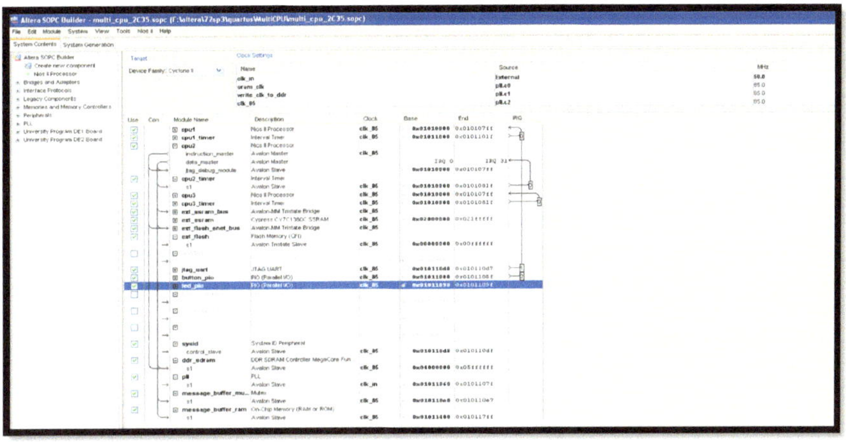

Multiprocessor Embedded System

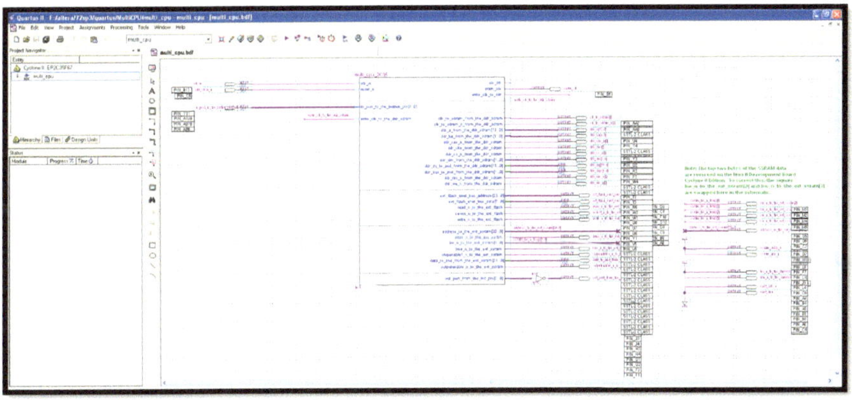

------------------C code to implement multiprocessor communication--------------

```
#include <stdio.h>
#include <string.h>
#include "sys/alt_alarm.h"
#include "system.h"
#include "nios2.h"
#include "altera_avalon_mutex.h"

#define MESSAGE_WAITING 1
#define NO_MESSAGE 0

#define LOCK_SUCCESS 0
#define LOCK_FAIL 1

#define MESSAGE_BUFFER_BASE MESSAGE_BUFFER_RAM_BASE

#define FIRST_LOCK 1 /* for testing only */
#define ERROR_OPENED_INVALID_MUTEX 1 /* for testing only */
#define ERROR_ALLOWED_ACCESS_WITHOUT_OWNING_MUTEX 2 /* for testing only */
#define ERROR_COULDNT_OPEN_MUTEX 3 /* for testing only */

#define MS_DELAY 1000

// Message buffer structure
typedef struct {
  char flag;
  char buf[100];
} message_buffer_struct;

int main()
{
  alt_mutex_dev* mutex = NULL;  // Pointer to our mutex device

  // Local variables
  unsigned int id;
  unsigned int value;
  unsigned int count = 0;
  unsigned int ticks_at_last_message;

  char got_first_lock = 0; /* for testing only */
  unsigned int error_code = 0; /* for testing only */
```

```
  message_buffer_struct *message;


  NIOS2_READ_CPUID(id);
  id += 1;

  value = 1;

  message = (message_buffer_struct*)MESSAGE_BUFFER_BASE;

  mutex = altera_avalon_mutex_open("/dev/wrong_device_name");   if (mutex !=
NULL)  {
        error_code = ERROR_OPENED_INVALID_MUTEX;
    goto error;  }

    mutex = altera_avalon_mutex_open(MESSAGE_BUFFER_MUTEX_NAME);

  ticks_at_last_message = alt_nticks();

  if (mutex)
  {

      if(altera_avalon_mutex_trylock(mutex, value) == LOCK_SUCCESS)      {
        if (altera_avalon_mutex_first_lock(mutex) == FIRST_LOCK)        {
          message->flag = NO_MESSAGE; /* for testing only */
          got_first_lock = 1; /* for testing only */
        }
        altera_avalon_mutex_unlock(mutex); /* for testing only */
      }

    while(1)
    {
      if (alt_nticks() >= (ticks_at_last_message + ((alt_ticks_per_second() *
(MS_DELAY)) / 1000)))
      {
        ticks_at_last_message = alt_nticks();

        // Try and aquire the mutex (non-blocking).
        if(altera_avalon_mutex_trylock(mutex, value) == LOCK_SUCCESS)
        {
          // Just make sure we own the mutex
          if(altera_avalon_mutex_is_mine(mutex)) /* for testing only */
          {
            // Check if the message buffer is empty
            if(message->flag == NO_MESSAGE)
            {
              count++;
              // If we were the first to lock the mutex, say so in our first
message.
              if (got_first_lock) /* for testing only */
              {
                sprintf(message->buf, "FIRST LOCK - Message from CPU %d.
Number sent: %d\n", id, count); /* for testing only */
                got_first_lock = 0; /* for testing only */
              }
              else
              {
                sprintf(message->buf, "Message from CPU %d.  Number sent:
%d\n", id, count);
              }
              // Set the flag that a message has been put in the buffer.
              message->flag = MESSAGE_WAITING;
```

```
            }
          }
          else            {
            error_code = ERROR_ALLOWED_ACCESS_WITHOUT_OWNING_MUTEX;
goto error; /* for testing only */
          }
          // Release the mutex
          altera_avalon_mutex_unlock(mutex);
        }
      }
#ifdef JTAG_UART_NAME
      {
        if(message->flag == MESSAGE_WAITING)
        {

          altera_avalon_mutex_lock(mutex, value);  /* for testing only */

          if(altera_avalon_mutex_is_mine(mutex)) /* for testing only */
          {
            printf("%s", message->buf);
            message->flag = NO_MESSAGE;
          }
          else            {
            error_code = ERROR_ALLOWED_ACCESS_WITHOUT_OWNING_MUTEX;
goto error;           }
                    altera_avalon_mutex_unlock(mutex); /* for testing only */
        }
      }
#endif
    }
  }
  else   {
    error_code = ERROR_COULDNT_OPEN_MUTEX; /* for testing only */
    goto error;  }

error:  return(error_code); }
```

6.5 Robotic ARM Controlled Over Ethernet

With an advent of considerable exponential growth of the Internet and all of its computing hardware/software technologies, it is possible to design a tele-operated robots controlled over the Internet. The operations like dangerous, hostile, and inaccessible to humans and areas of work regardless of geographical locations Internet-based tele-operated robot finds great utility in such situations. The Internet being so matured and freely available, people can get connected and allow access to devices across the globe.

Small robotic arm has been designed which is operated over the LAN using the TCP/IP protocol. VB.NET is used to design a graphical user interface (GUI) client application which sends control instructions over a LAN using TCP/IP. Remote server receives these instructions, decodes it, and moves the motors of the robotic arm accordingly.

The designed robotic arm is of 2 degrees of freedom (DOF) which is able to turn left, right, down, and up. Figure 6.6 illustrates robotic arm control setup.

The main block diagram of entire system is shown in Fig. 6.7.

.NET framework needs to be installed on the client side so that the GUI works. The main reason behind using VB.NET is that it simplifies the process of designing GUI applications.

The server is implemented on an Altera DE2 development board which has Cyclone® II 2C35 FPGA on which we implement a customized Nios II configurable processor. Socket communication on the DE2 board is established by loading lightweight IP, TCP/IP stack on Nios II. DM9000A Ethernet PHY/MAC controller is already present on the board which simplifies the Ethernet interface.

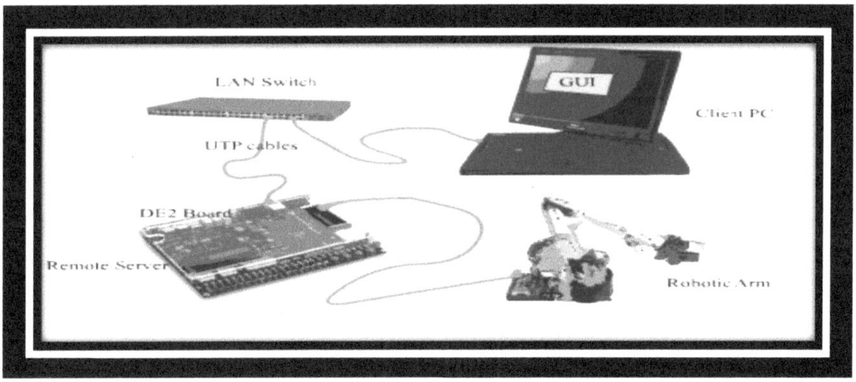

Fig. 6.6 Block of robotic arm control

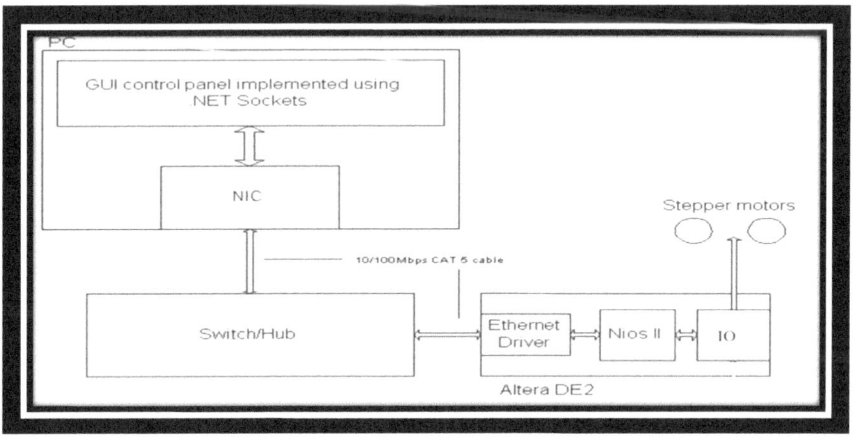

Fig. 6.7 Detailed block of robotic arm control

Design of the Robotic Arm

Here, we have used two stepper motors to design a robotic arm. X-axis rotation is controlled by one stepper. The second stepper motor controls the rotation along the Y-axis.

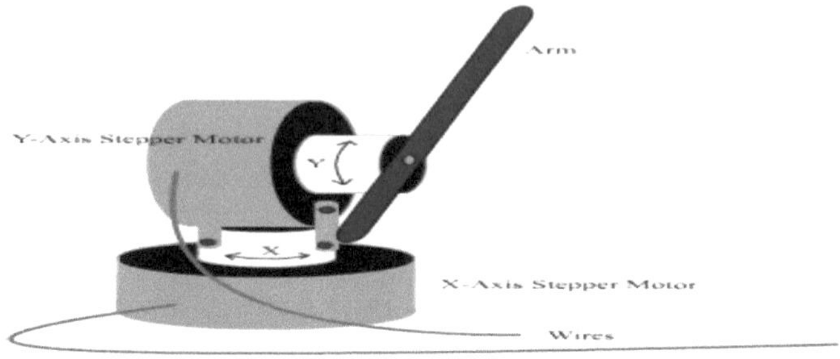

Client Interface

The GUI was designed in Microsoft Visual Studio 2008 and coded using VB.NET.

The GUI acts as a client that tries to connect to the server that is programmed onto the DE2 board. The GUI is used to control the stepper motors of the robotic arm by transmitting control messages. The sliders on the home tab page of the GUI are used to specify the number of steps and direction that the stepper motor should move.

The GUI uses the **system.net.sockets** and **system.net** namespaces of the .NET framework. The GUI uses the socket to act as a TCP/IP client.

In order to have a smoothly operating GUI, we also required to use some multithreading concepts. This was used mainly to update the interface while trying to maintain the TCP connections. The updating of the controls on the GUI was sometimes required to be run on a separate thread.

The various features of the GUI will be discussed with reference to Fig. 6.8.

1. There are three tab pages named Home, Ping, and Network Info.
   (a) Home: This is the main page that you use to control the stepper motor. This is the page that allows you to connect to the server.
   (b) Ping: This page provides a simple utility that allows you to ping to a remote host. This saves us the trouble of pinging from DOS.
   (c) Network Info: This tab page gives the basic information on the available network connections present on the host machine on which the client GUI is running.

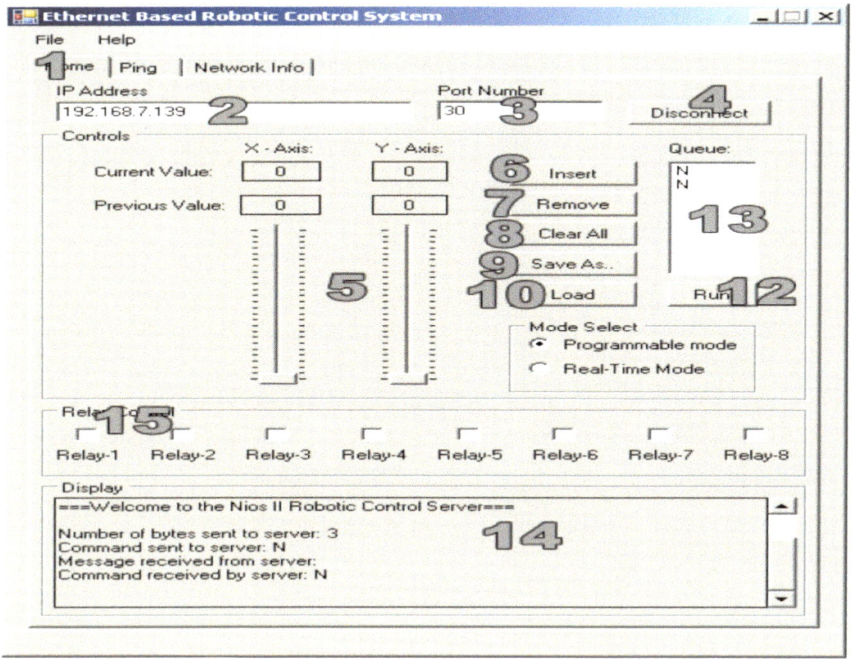

Fig. 6.8 GUI for robotic arm control

2. This is where you enter the IP address of the server that is programmed onto the DE2 board.
3. This is where you enter the port number. This should match that of the server socket port number.
4. The Connect button is used to connect to or disconnect from the server.
5. These are sliders that can be moved with either the mouse or the keyboard arrow keys. The value on the slider specifies the number of steps for the stepper motor. Moving the slider up will result in forward motion, and moving the slider down will reverse the motors' movement. The left slider controls the left/right movement of the robotic arm. The slider toward the right specifies the up/down movement of the robotic arm.
6. The Insert button is used to insert the steps in a queue. This can be used to program the robotic arms' movements.
7. The Remove button is used to remove values from the queue.
8. The Clear All button clears the steps from the queue.
9. The Save As button is used to save the queue entries to an external text file for later use.
10. The Load button is used to load previously saved queue entries.

11. The GUI has two modes.
    (a) Programmable mode: In this mode, the user has to first fill up the queue and then hit the Run button to send the instructions to the robotic arm. He has no real-time control over the arm.
    (b) Real-time mode: In this mode, the user has real-time control over the robotic arm. He can send only one instruction at a time.
12. The Run button is used to send the entries from the queue as packets to the server.
13. This is the queue that holds the step entries.
14. This displays the status of the socket connection and the number of bytes sent to the server.
15. These checkboxes are used to toggle the relays connected to the expansion header of the DE2 board.

The Remote Server

The remote server is implemented on an Altera DE2 development board that has Cyclone® II 2C35 FPGA on which we implement a Nios® II configurable processor. The board has a DM9000A 10/100 Ethernet PHY/MAC controller.

To implement the server on Nios® II, we first needed to embed a TCP/IP stack onto the Nios® II soft processor. We choose the lightweight IP (LwIP) TCP/IP stack. Altera platform supports LwIP along with MicroC/OS-II RTOS multi-threaded environment. Therefore, to use lwIP, you must base your C/C++ project on the MicroC/OS-II RTOS.

LwIP

Adam Dunkels from Computer and Networks Architectures (CNA) laboratory at the Swedish Institute of Computer Science (SICS) has developed Independent LwIP which is a small version of TCP/IP protocol.

The main intention behind using LwIP stack is to reduce memory usage and code size, making lwIP suitable for use in small clients with very limited resources such as embedded systems. LwIP uses a tailor-made API in order to reduce processing and memory demands.

The DE2 board receives these packets through its Ethernet port and sends it to the Nios® II processor where the packet gets decoded. The processor then accordingly sends control bits to the board expansion headers where the stepper motor driver circuit is connected.

The C Code for the Entire Nios II System to Control Robotic Arm is Given Below

Code for web_server.c

```
#include <stdio.h>
#include <errno.h>
#include <ctype.h>
#include "includes.h"
#include "alt_lwip_dev.h"
#include "lwip/sys.h"
#include "user.h"
#include "alt_error_handler.h"
#include "altera_avalon_pio_regs.h"
#include "dm9000.h"
#include "lcd.h"
ALTERA_AVALON_DM9K_INSTANCE(DM9000A, dm9k);
void user_task(void * pvoid)
{
static u_long val=0xF0;
// simply doing sanity check
for(;;)
{
val = ~val;
IOWR_ALTERA_AVALON_PIO_DATA(LED_GREEN_BASE, val);
usleep(1000000);
} // of forever loop
}
#ifndef LWIP
#error This Server requires the Lightweight IP Software Component.
#endif
#ifndef __ucosii__
#error This Server requires the UCOS II IP Software Component.
#endif
OS_EVENT *attained_ip_address_sem;
static void tcpip_init_done(void *arg)
{
// hychu
ALTERA_AVALON_DM9K_INIT(dm9k);

if (!lwip_devices_init(ETHER_PRIO))
die_with_error("[tcpip_init_done] Fatal: Can't add ethernet interface!");
attained_ip_address_sem = OSSemCreate(1);
#if LWIP_DHCP == 1
if(!(IORD(SWITCH_PIO_BASE, 0) & (1<<17))) sys_thread_new(dhcp_timeout_task,
NULL, DHCP_TMR_PRIO);
#endif /* LWIP_DHCP */

if(!sys_thread_new(http_task, NULL, HTTP_PRIO))
die_with_error("[tcpip_init_done] Fatal: Can't add HTTP task! aka SERVER
TASK");
}
int main ()
{
INT8U error_code;
LCD_Init();
lwip_stack_init(TCPIP_PRIO, tcpip_init_done, 0);

error_code = OSTaskCreateExt(SSSInitialTask,
NULL,
(void *)&SSSInitialTaskStk[TASK_STACKSIZE],
SSS_INITIAL_TASK_PRIORITY,
SSS_INITIAL_TASK_PRIORITY,
SSSInitialTaskStk,
TASK_STACKSIZE,
```

```
NULL,
0);
alt_uCOSIIErrorHandler(error_code, 0);
printf("\nThe Nios II Robotic Control Server is starting up\n");
// hychu
sys_thread_new(user_task, NULL, SANITY_PRIO);
OSStart();
return 0;
}
```

Code for simple_socket_server.c

```
#include <stdio.h>
#include <string.h>
#include <ctype.h>
#include "includes.h"
#include "alt_error_handler.h"
#include "altera_avalon_pio_regs.h"
#include "alt_lwip_dev.h"
#include "lwip/sys.h"
#include "lwip/netif.h"
#include "lwip/sockets.h"
#include "user.h"
/* Local Function Prototypes */
void SSSCreateOSDataStructs();
void SSSCreateTasks();
OS_EVENT *SSSLEDCommandQ;
#define SSS_LED_COMMAND_Q_SIZE 100
void *SSSLEDCommandQTbl[SSS_LED_COMMAND_Q_SIZE]; /*Storage for
OS_FLAG_GRP *SSSLEDEventFlag;
OS_EVENT *SSSLEDLightshowSem;
OS_STK SSSInitialTaskStk[TASK_STACKSIZE];

OS_STK LEDManagementTaskStk[TASK_STACKSIZE];
OS_STK LED7SegLightshowTaskStk[TASK_STACKSIZE];
void SSSInitialTask(void* pdata)
{
INT8U error_code = OS_NO_ERR;
/*create os data structures */
SSSCreateOSDataStructs();
error_code = OSTaskDel(OS_PRIO_SELF);
alt_uCOSIIErrorHandler(error_code, 0);
while (1);
void SSSCreateOSDataStructs(void)
{
INT8U error_code;
SSSLEDCommandQ = OSQCreate(&SSSLEDCommandQTbl[0], SSS_LED_COMMAND_Q_SIZE);
if (!SSSLEDCommandQ)
{
alt_uCOSIIErrorHandler(EXPANDED_DIAGNOSIS_CODE,
"Failed to create SSSLEDCommandQ.\n");
}

SSSLEDLightshowSem = OSSemCreate(1);
if (!SSSLEDLightshowSem)
{
alt_uCOSIIErrorHandler(EXPANDED_DIAGNOSIS_CODE,
"Failed to create SSSLEDLightshowSem.\n");

SSSLEDEventFlag = OSFlagCreate(0, &error_code);
if (!SSSLEDEventFlag)
```

```
{
alt_uCOSIIErrorHandler(error_code, 0);
}

attained_ip_address_sem = OSSemCreate(0);
if (!attained_ip_address_sem)
{
alt_uCOSIIErrorHandler(EXPANDED_DIAGNOSIS_CODE,
"Failed to create attained_ip_address_sem.\n");
}
}
void SSSCreateTasks(void)
{
void sss_reset_connection(SSSConn* conn)
{
memset(conn, 0, sizeof(SSSConn));
conn->fd = -1;
conn->state = READY;
conn->rx_wr_pos = conn->rx_buffer;
conn->rx_rd_pos = conn->rx_buffer;
return;
}

void sss_send_menu(SSSConn* conn)
{
alt_u8 tx_buf[SSS_TX_BUF_SIZE];
alt_u8 *tx_wr_pos = tx_buf;
tx_wr_pos += sprintf(tx_wr_pos,"\r\n===Welcome to the Nios II Robotic Control
Server===\r");
send(conn->fd, tx_buf, tx_wr_pos - tx_buf, 0);
return;
}
void sss_handle_accept(int listen_socket, SSSConn* conn)
{
int socket, len;
struct sockaddr_in incoming_addr;
len = sizeof(incoming_addr);
if ((conn)->fd == -1)
{
if((socket=accept(listen_socket,(struct sockaddr*)&incoming_addr,&len))<0)


alt_lwIPErrorHandler(EXPANDED_DIAGNOSIS_CODE,
"[sss_handle_accept] accept failed");
}
else
{
(conn)->fd = socket;
sss_send_menu(conn);
printf("[sss_handle_accept] accepted connection request from %s\n",
inet_ntoa(incoming_addr.sin_addr));
}
}
else
{
printf("[sss_handle_accept] rejected connection request from %s\n",
inet_ntoa(incoming_addr.sin_addr));
}
return;
}
void sss_exec_command(SSSConn* conn)
{
```

```
int bytes_to_process = conn->rx_wr_pos - conn->rx_rd_pos;
INT8U tx_buf[SSS_TX_BUF_SIZE];
INT8U *tx_wr_pos = tx_buf;
// INT8U error_code;
int j=0;
alt_u8 step[4] = {0xA,0x6,0x5,0x9};

INT8U SSSCommand;
SSSCommand = CMD_LEDS_BIT_0_TOGGLE;
while(bytes_to_process--)
{
SSSCommand = toupper(*(conn->rx_rd_pos++));
if(SSSCommand >= ' ' && SSSCommand <= '~')
{
tx_wr_pos += sprintf(tx_wr_pos,
"\r\nCommand received by server: %c\r",
SSSCommand);
if (SSSCommand == CMD_QUIT)
{
tx_wr_pos += sprintf(tx_wr_pos,"\r\n===Terminating connection===\n\r");
conn->close = 1;
}
else
{

#ifdef LED_RED_BASE
static int up1 = 0;
static int down1 = 3;
static int left1 = 0;
static int right1 = 3;
static int toggle1 = 0;
static int toggle2 = 0;
static int toggle3 = 0;
static int toggle4 = 0;
static int toggle5 = 0;
static int toggle6 = 0;
static int toggle7 = 0;
static int toggle8 = 0;
static int relayval = 0;
switch (SSSCommand) {
case 'U':
if (up1 == 4)
{
up1 = 0;
}
IOWR_ALTERA_AVALON_PIO_DATA(MOTORLED_2_BASE, step[up1]);
IOWR_ALTERA_AVALON_PIO_DATA(MOTOR2_BASE, step[up1]);
printf("up1 value = %i\n", up1);
up1++;
down1 = up1 - 2;
usleep(90000);
printf("Move motor one step UP\n");
break;
case 'D':
if (down1 == -1)
{
down1 = 3;
}
IOWR_ALTERA_AVALON_PIO_DATA(MOTORLED_2_BASE, step[down1]);
IOWR_ALTERA_AVALON_PIO_DATA(MOTOR2_BASE, step[down1]);
printf("down1 value = %i\n", down1);
down1--;
```

```
up1 = down1 + 2;

usleep(90000);
printf("Move motor one step DOWN\n");
break;
case 'L':
if (left1 == 4)
{
left1 = 0;
}
IOWR_ALTERA_AVALON_PIO_DATA(MOTORLED_1_BASE, step[left1]);
IOWR_ALTERA_AVALON_PIO_DATA(MOTOR1_BASE, step[left1]);
printf("left1 value = %i\n", left1);
left1++;
right1 = left1 - 2;
usleep(90000);
printf("Move motor one step LEFT\n");
break;
case 'R':
if (right1 == -1)
{
right1 = 3;
}
IOWR_ALTERA_AVALON_PIO_DATA(MOTORLED_1_BASE, step[right1]);
IOWR_ALTERA_AVALON_PIO_DATA(MOTOR1_BASE, step[right1]);
printf("right1 value = %i\n", right1);
right1--;
left1 = right1 + 2;
usleep(90000);
printf("Move motor one step RIGHT\n");
break;
case '1':
toggle1 = ~toggle1;
if(toggle1)
{
relayval = relayval | 0x01;
IOWR_ALTERA_AVALON_PIO_DATA(RELAY_BASE, relayval);
IOWR_ALTERA_AVALON_PIO_DATA(RELAYLED_BASE, relayval);
}
else
{
relayval = relayval & 0xfe;
IOWR_ALTERA_AVALON_PIO_DATA(RELAY_BASE, relayval);
IOWR_ALTERA_AVALON_PIO_DATA(RELAYLED_BASE, relayval);
}
break;
case '2':
toggle2 = ~toggle2;
if(toggle2)
{
relayval = relayval | 0x02;
IOWR_ALTERA_AVALON_PIO_DATA(RELAY_BASE, relayval);
IOWR_ALTERA_AVALON_PIO_DATA(RELAYLED_BASE, relayval);
}
else
{
relayval = relayval & 0xfd;
IOWR_ALTERA_AVALON_PIO_DATA(RELAY_BASE, relayval);
IOWR_ALTERA_AVALON_PIO_DATA(RELAYLED_BASE, relayval);
}
break;
case '3':
```

```
toggle3 = ~toggle3;
if(toggle3)
{
relayval = relayval | 0x04;
IOWR_ALTERA_AVALON_PIO_DATA(RELAY_BASE, relayval);
IOWR_ALTERA_AVALON_PIO_DATA(RELAYLED_BASE, relayval);
}
else
{
relayval = relayval & 0xfb;
IOWR_ALTERA_AVALON_PIO_DATA(RELAY_BASE, relayval);
IOWR_ALTERA_AVALON_PIO_DATA(RELAYLED_BASE, relayval);
}
break;
case '4':
toggle4 = ~toggle4;
if(toggle4)
{
relayval = relayval | 0x08;
IOWR_ALTERA_AVALON_PIO_DATA(RELAY_BASE, relayval);
IOWR_ALTERA_AVALON_PIO_DATA(RELAYLED_BASE, relayval);
}
else
{
relayval = relayval & 0xf7;
IOWR_ALTERA_AVALON_PIO_DATA(RELAY_BASE, relayval);
IOWR_ALTERA_AVALON_PIO_DATA(RELAYLED_BASE, relayval);
}
break;
case '5':
toggle5 = ~toggle5;
if(toggle5)
{
relayval = relayval | 0x10;
IOWR_ALTERA_AVALON_PIO_DATA(RELAY_BASE, relayval);
IOWR_ALTERA_AVALON_PIO_DATA(RELAYLED_BASE, relayval);
}
else
{
relayval = relayval & 0xef;
IOWR_ALTERA_AVALON_PIO_DATA(RELAY_BASE, relayval);
IOWR_ALTERA_AVALON_PIO_DATA(RELAYLED_BASE, relayval);
}
break;
case '6':
toggle6 = ~toggle6;
if(toggle6)
{
relayval = relayval | 0x20;
IOWR_ALTERA_AVALON_PIO_DATA(RELAY_BASE, relayval);
IOWR_ALTERA_AVALON_PIO_DATA(RELAYLED_BASE, relayval);
}
else
{
relayval = relayval & 0xdf;
IOWR_ALTERA_AVALON_PIO_DATA(RELAY_BASE, relayval);
IOWR_ALTERA_AVALON_PIO_DATA(RELAYLED_BASE, relayval);
}
break;
case '7':
toggle7 = ~toggle7;
if(toggle7)
```

```
{
relayval = relayval | 0x40;
IOWR_ALTERA_AVALON_PIO_DATA(RELAY_BASE, relayval);
IOWR_ALTERA_AVALON_PIO_DATA(RELAYLED_BASE, relayval);
}
else
{
relayval = relayval & 0xbf;
IOWR_ALTERA_AVALON_PIO_DATA(RELAY_BASE, relayval);
IOWR_ALTERA_AVALON_PIO_DATA(RELAYLED_BASE, relayval);
}
break;
case '8':
toggle8 = ~toggle8;
if(toggle8)
{
relayval = relayval | 0x80;
IOWR_ALTERA_AVALON_PIO_DATA(RELAY_BASE, relayval);
IOWR_ALTERA_AVALON_PIO_DATA(RELAYLED_BASE, relayval);
}
else
{
relayval = relayval & 0x7f;
IOWR_ALTERA_AVALON_PIO_DATA(RELAY_BASE, relayval);
IOWR_ALTERA_AVALON_PIO_DATA(RELAYLED_BASE, relayval);
}
break;
default:
IOWR_ALTERA_AVALON_PIO_DATA(MOTORLED_1_BASE, 0x00);
IOWR_ALTERA_AVALON_PIO_DATA(MOTORLED_2_BASE, 0x00);
printf("STALL MOTOR for 1 second\n");
usleep(90000);
break;
}
#endif
}
}
}
send(conn->fd, tx_buf, tx_wr_pos - tx_buf, 0);
return;
}
void sss_handle_receive(SSSConn* conn)
{
int data_used = 0, rx_code = 0;
INT8U *lf_addr;
conn->rx_rd_pos = conn->rx_buffer;
conn->rx_wr_pos = conn->rx_buffer;
printf("[sss_handle_receive] processing RX data\n");
while(conn->state != CLOSE)
{
lf_addr = strchr(conn->rx_buffer, '\n');
if(lf_addr)
{
sss_exec_command(conn);
}
else
{
rx_code = recv(conn->fd, conn->rx_wr_pos,
SSS_RX_BUF_SIZE - (conn->rx_wr_pos - conn->rx_buffer) -1, 0);
if(rx_code > 0)
{
conn->rx_wr_pos += rx_code;
```

```
*(conn->rx_wr_pos+1) = 0;
}
}

conn->state = conn->close ? CLOSE : READY;
data_used = conn->rx_rd_pos - conn->rx_buffer;
memmove(conn->rx_buffer, conn->rx_rd_pos,
conn->rx_wr_pos - conn->rx_rd_pos);
conn->rx_rd_pos = conn->rx_buffer;
conn->rx_wr_pos -= data_used;
memset(conn->rx_wr_pos, 0, data_used);
}
printf("[sss_handle_receive] closing connection\n");
close(conn->fd);
sss_reset_connection(conn);
return;
}

void http_task()
{
int fd_listen, max_socket;
struct sockaddr_in addr;
static SSSConn conn;
fd_set readfds;
if ((fd_listen = socket(AF_INET, SOCK_STREAM, 0)) < 0)
{
alt_lwIPErrorHandler(EXPANDED_DIAGNOSIS_CODE,"[sss_task] Socket creation
failed");
}
addr.sin_family = AF_INET;
addr.sin_port = htons(SSS_PORT);
addr.sin_addr.s_addr = INADDR_ANY;
if ((bind(fd_listen,(struct sockaddr *)&addr,sizeof(addr))) < 0)
{
alt_lwIPErrorHandler(EXPANDED_DIAGNOSIS_CODE,"[sss_task] Bind failed");
}

if ((listen(fd_listen,1)) < 0)
{
alt_lwIPErrorHandler(EXPANDED_DIAGNOSIS_CODE,"[sss_task] Listen failed");
}

sss_reset_connection(&conn);
printf("[sss_task] Simple Socket Server listening on port %d\n", SSS_PORT);
while(1)
{
FD_ZERO(&readfds);
FD_SET(fd_listen, &readfds);

max_socket = fd_listen+1;
if (conn.fd != -1)
{
FD_SET(conn.fd, &readfds);
if (max_socket <= conn.fd)
{
max_socket = conn.fd+1;
}
}
select(max_socket, &readfds, NULL, NULL, NULL);
if (FD_ISSET(fd_listen, &readfds))
{
sss_handle_accept(fd_listen, &conn);
```

```
}
else
{
if ((conn.fd != -1) && FD_ISSET(conn.fd, &readfds))
{
sss_handle_receive(&conn);
}
}
}
}
```

The remaining codes for the following utilities are not provided here; if somebody wants these, a personnel request may be sent to authors.

Code for network_utilities.c
Code for user.h
Code for alt_error_handler.c
Code for dm9000.c

6.6 Multivariate System Implementation

Current advancement in computational technology and instrumentation techniques enables us to collect and process huge amounts of data from chemical and biological processes. Multivariate Statistical Process Control (MVSPC) has become very popular tool to extract the useful information from the measured input data for improving the product quality process and performance. During the last several years, it has been successfully applied and tested for monitoring and modeling of chemical and biological processes.

One of the most popular MVSPC techniques is partial least squares (PLS). PLS is a multivariate process identification method that projects the input–output data down into a latent space, extracting a number of principal factors with an orthogonal structure, while capturing most of the variance in the original data.

SOPC Components for NIOS II System Creation

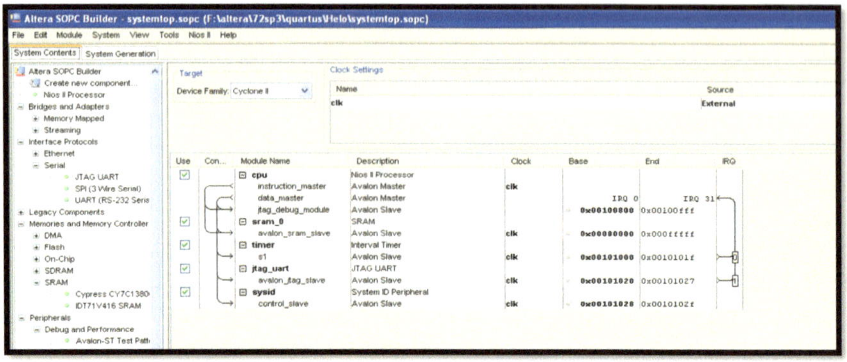

Here, the system generated is very simple, and the components required are Main Nios II cpu, SRAM, Interval time, JTAG UART, and System ID.

Full-Fledged Nios II System for Multivariate Analysis

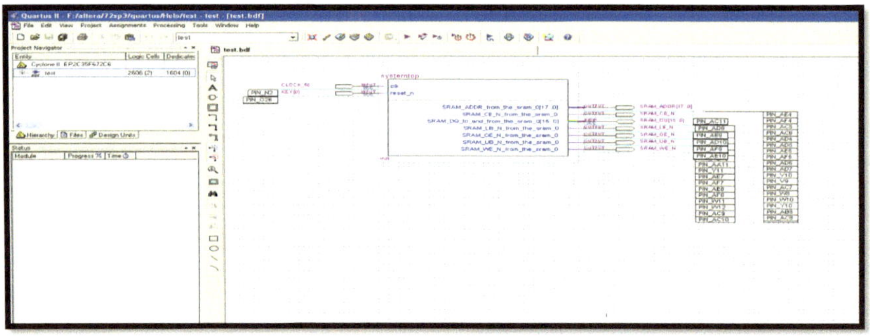

Compilation Report

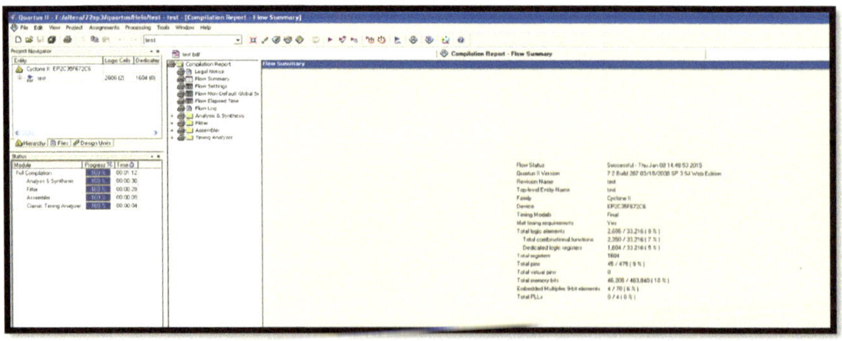

C Code for Multivariate Partial Least Square Regression

```
#include <stdio.h>
#include <math.h>
#include <float.h>
#define ROW1 4
#define COL1 4
#define ROW2 4
#define COL2 4
#define N 4

double** minus1(double** ,double**,int );
double addition1(double** ,int ,int);
double** wpq1(double** ,int ,int);
double**divide1(double**, int,int,double);
double** column(double**,int,int);
void Jacobi1_Cyclic1_Method1(double eigenvalues1[COL1], double
*eigenvectors1[COL1][COL1],double *A, int n);
double** identity(int);
double** transpose1(double**,int,int);
double** init1(double**,int,int);
double** set1(double**,int,int);
void get(double**,int,int);
double** mul1(double**,double**,int,int,int);
void main()
{
int i,j
double**matrix1,**matrix2,**AO,**MO,**trans1,**CO,**AO_trans1,**g,M[COL1][COL
1];
    double eigenvalues1[COL1],**qh,**Wh,**Wh_mat,**Ch,**W,**ph,**p,**q,**vh;
    double eigenvectors1[COL1][COL1],Ch_sq,m=2.0,**X_pre;
    double  **v_trans1,**C1,**p_trans1,**M1,**A1,**q_trans1,**B,**T,av_vh;
    clrscr();
matrix1=init(matrix1,ROW1,COL1);
    matrix2=init(matrix2,COL2,COL2);

     set1(matrix1,ROW1,COL1);
 set1(matrix2,ROW2,COL2);
clrscr();
trans1=transpose1(matrix1,COL1,ROW1);
AO=mul1(trans,matrix2,COL1,COL2,ROW1);
MO=mul1(trans,matrix1,COL1,COL1,ROW1);
CO=identity(COL1);
AO_trans1=transpose1(AO,COL2,COL1);
g=mul(AO_trans1,AO,COL2,COL2,COL1);
    for(i=0;i<COL1;i++)
     {
      for(j=0;j<COL1;j++)
      {
       M[i][j]=*(*(g+i)+j);
      }
     }


Jacobi1_Cyclic1_Method1(eigenvalues1,*eigenvectors1,*M,COL1);
qh=init1(qh,COL1,COL1);
 for(i=0;i<COL1;i++)
    {
     for(j=0;j<COL1;j++)
     {
      if(i==j)
       qh[i][j]=eigenvalues1[i];
      else
       qh[i][j]=0.0;
```

```
        }
       }
        Wh=mul1(AO,qh,COL1,COL1,COL2);
        Wh_mat1=column(Wh,COL1,COL1);
        Ch=transpose1(Wh_mat,1,COL1);
        Ch=mul1(Ch,MO,1,COL1,COL1);
        Ch=mul1(Ch,Wh_mat,1,1,COL1);
        Ch_sq=sqroot(**Ch);
        Wh_mat1=divide1(Wh_mat,COL1,1,Ch_sq);
        W=wpq1(Wh_mat,COL1,1);

        Wh_ma1=column(W,COL1,1);
        ph=mul1(MO,Wh_mat,COL1,1,COL1);
        p=wpq1(ph,COL1,1);

    Wh_mat=column(W,COL1,1);
        qh=mul1(AO_trans1,Wh_mat,COL2,1,COL1);
    q=wpq1(qh,COL2,1);
        ph=column(p,COL1,1);
        vh=mul1(CO,ph,COL1,1,COL1);
        av_vh=addition1(vh,COL1,1);
        av_vh=av_vh/m;
        vh=divide1(vh,COL1,1,av_vh);
        v_trans1=transpose1(vh,1,COL1);
       C1=mul1(vh,v_trans,COL1,COL1,1);
       C1=minus1(CO,C1,COL1);
        p_trans1=transpose1(ph,1,COL1);
     M1=mul1(ph,p_trans1,COL1,COL1,1);
         M1=minus1(MO,M1,COL1);
     A1=mul1(CO,AO,COL1,COL2,COL1);
         q_trans1=transpose1(q,1,COL1);
         B=mul1(W,q_trans1,COL1,COL1,1);
         T=mul1(matrix1,W,ROW1,1,COL1);
         get(T,ROW1,1);
          ph=transpose1(p,1,COL1);
          X_pre=mul1(T,ph,ROW1,COL1,1);
          matrix1=transpose1(matrix1,COL1,ROW1);
          X_pre=transpose1(X_pre,COL1,ROW1);
          get(X_pre,COL1,ROW1);

       getch();
       free(matrix1);
       free(matrix2);

    } /* end main */


    double** init1(double** arr1,int row,int col)
    {
     int i=0,j=0;
     arr1=(double**)malloc(sizeof(double)*row*col);
          for(i=0;i<row;i++)
          {
                  for(j=0;j<col;j++)
                  {
                   *((arr1+i)+j)=(double*)malloc(sizeof(double));
                   *(*(arr1+i)+j)=0.0;
                  }
          }
     return arr1;
    }
```

```
double** set1(double** arr1,int row,int col)
{
 int i=0,j=0;
 double val=0.0;
 for(i=0;i<row;i++)
      {
              for(j=0;j<col;j++)
              {
printf("Enter value of row %d col %d  :",(i+1),(j+1));
          scanf("%lf",&val);
          *(*(arr1+i)+j)=val;
              }
      }
      return arr1;
}

void get(double** arr1,int row,int col)
{
      int i=0,j=0;

      for(i=0;i<row;i++)
      {
              for(j=0;j<col;j++)
              {
               printf("%lf\t",*(*(arr1+i)+j));
              }
            printf("\n");
      }
}
double** mul1(double** arr2,double** arr3,int row,int col,int col1)
{
 double  **result;
 int i=0,j=0,k=0;
 result=init1(result,row,col);
   for(i=0;i<row;i++)
      {
              for(j=0;j<col;j++)
              {
                      for(k=0;k<col1;k++)
                      {
                         *(*(result+i)+j)+=(*(*(arr2+i)+k))*(*(*(arr3+k)+j));
                         if (k!=(col1-1))
                         printf("+");
                      }
                 printf("\t");
              }
       printf("\n");
      }
 return (result);
}

double** transpose1(double** arr2,int row1,int col1)
{
double **trans;
int i,j;
trans1=init(trans1,row1,col1);
for(i=0;i<col1;i++)
{
 for(j=0;j<row1;j++)
 *(*(trans1+j)+i)=*(*(arr1+i)+j);
}
return trans1;
```

```
}

double** identity(int dim1)
{
double **CO;
int i,j;
CO=init(CO,dim1,dim1);
for(i=0;i<dim1;i++)
{
 for(j=0;j<dim1;j++)
 {
 if(i==j)
 {
 CO[i][j]=1.0;
 }
 else
 {
 CO[i][j]=0.0;
 } }}
return CO;
}

Void Jacobi1_Cyclic1_Method1 (double eigenvalues1[N],
double*eigenvectors1[N][N],double *A, int n)
{
int row, i, j, k, m;
double *pAk, *pAm, *p_r, *p_e;
double threshold_norm;
double threshold;
double tan1_phi, sin_phi, cos_phi, tan2_phi, sin1_phi, cos1_phi;
double sin_2phi, cos_2phi, cot_2phi;
double dum1;
double dum2;
double dum3;
double r;
double max;
  if ( n < 1) return;
  if ( n == 1) {
    eigenvalues1[0] = *A;
   *eigenvectors1[0][0] = 1.0;
     return;
  }

  for (p_e = eigenvectors1, i = 0; i < n; i++)
     for (j = 0; j < n; p_e++, j++)
        if (i == j)
        *p_e = 1.0; else *p_e = 0.0;
 for (threshold = 0.0, pAk1 = A, i = 0; i < ( n - 1 ); pAk1 += n, i++)
     for (j = i + 1; j < n; j++) threshold1 += *(pAk1 + j) * *(pAk1 + j);
  threshold1 = sqroot(threshold1 + threshold1);
  threshold_norm = threshold1 * DBL_EPSILON;
  max = threshold1 + 1.0;
  while (threshold1> threshold_norm) {
     threshold1 /= 10.0;
     if (max < threshold1) continue;
     max = 0.0;
     for (pAk1 = A, k = 0; k < (n-1); pAk1 += n, k++) {
for (pAm1 = pAk1 + n, m = k + 1; m < n; pAm1 += n, m++) {
  if ( fabs(*(pAk1 + m)) < threshold ) continue;
cot_2phi = 0.5 * ( *(pAk1 + k) - *(pAm1 + m) ) / *(pAk1 + m);
           dum1 = sqroot( cot_2phi * cot_2phi + 1.0);
           if (cot_2phi < 0.0) dum1 = -dum1;
```

```
            tan_phi = -cot_2phi + dum1;
            tan2_phi = tan_phi * tan_phi;
            sin2_phi = tan2_phi / (1.0 + tan2_phi);
            cos2_phi = 1.0 - sin2_phi;
            sin_phi = sqroot (sin2_phi);
            if (tan_phi < 0.0) sin_phi = - sin_phi;
            cos_phi = sqroot (cos2_phi);
            sin_2phi = 2.0 * sin_phi * cos_phi;
            cos_2phi = cos2_phi - sin2_phi;
            p_r = A;
            dum1 = *(pAk1 + k);
            dum2 = *(pAm1 + m);
            dum3 = *(pAk1 + m);
            *(pAk + k) = dum1 * cos2_phi + dum2 * sin2_phi + dum3 * sin_2phi;
            *(pAm + m) = dum1 * sin2_phi + dum2 * cos2_phi - dum3 * sin_2phi;
            *(pAk + m) = 0.0;
            *(pAm + k) = 0.0;
            for (i = 0; i < n; p_r += n, i++) {
               if ( (i == k) || (i == m) ) continue;
               if ( i < k ) dum1 = *(p_r + k);
                else
                  dum1 = *(pAk1 + i);
               if ( i < m ) dum2 = *(p_r + m); else dum2 = *(pAm + i);
               dum3 = dum1 * cos_phi + dum2 * sin_phi;
               if ( i < k ) *(p_r + k) = dum3; else *(pAk1 + i) = dum3;
               dum3 = - dum1 * sin_phi + dum2 * cos_phi;
               if ( i < m ) *(p_r + m) = dum3; else *(pAm1 + i) = dum3;
            }
            for (p_e = eigenvectors1, i = 0; i < n; p_e += n, i++) {
               dum1 = *(p_e + k);
               dum2 = *(p_e + m);
               *(p_e + k) = dum1 * cos_phi + dum2 * sin_phi;
               *(p_e + m) = - dum1 * sin_phi + dum2 * cos_phi;
            }
         }
         for (i = 0; i < n; i++)
            if ( i == k ) continue;
            else if ( max < fabs(*(pAk1 + i))) max = fabs(*(pAk1 + i));
      }
   }
   for (pAk1 = A, k = 0; k < n; pAk1 += n, k++) eigenvalues1[k] = *(pAk1 + k);
}
double** column(double** matrix,int row,int col)
{
int i,j,k=0;
double **column;
column=init(column,row,col);
for(i=0,j=(col-1);i<row;i++)
{
*(*(column+i)+k)=*(*(matrix+i)+j);
}
return column;
}
double** divide1(double** matrix,int row,int col,double Ch_sq)
{
int i,j,k=0;
double **divide1;
divide=init(column,row,col);
for(i=0,j=(col-1);i<row;i++)
{
*(*(divide1+i)+k)=*(*(matrix+i)+j) / Ch_sq;
}
```

```
return divide1;
}
double** wpq1(double** matrix,int row,int col)
{
int i,j,k=0;
double **wpq1;
wpq=init(wpq1,row,col);
for(i=0;i<row;i++)
{
*(*(wpq+i)+k)= *(*(matrix+i)+k);
}
return wpq1;
}
double addition1(double** matrix,int row,int col)
{
int i,j=col-1;
double add=0.0;
for(i=0;i<row;i++)
add1+= *(*(matrix+i)+j);
return add1;
}
double** minusmat(double** matrix1,double** matrix2,int col)
{
int i,j;
double **minusmat;
minusmat=init(minusmat,col,col);
for(i=0;i<col;i++)
{
 for(j=0;j<col;j++)
 *(*(minusmat+i)+j)=( *(*(matrix1+i)+j) - *(*(matrix2+i)+j) );
}
return minusmat;
}
```

6.7 Matrix Crunching on Altera DE2 Board

FPGA Platform have become a preferred choice over the others especially when the design application involves hardware implementation of highly computive algorithms,high performance, reconfigurability and time to market. Sophisticated algorithms involving kernel operation such as Matrix multiplication, transpose, Inverse which are normally used in applications like image, signal processing and communication can now be achieved using low cost FPGA platform instead of using dedicated multi processor system.

Nios II system for the implementation of matrix crunching problems such as matrix multiplication, matrix transpose, matrix inverse is same as that of multivariate system implementation where only the algorithms change.

- Matrix Multiplication

```
#include<stdio.h>

int main() {
   int amat[5][5], bmat[5][5], cmat[5][5], i, j, k;
   int sum1 = 0;

   printf("\nEnter First Matrix : n");
   for (i = 0; i < 3; i++) {
      for (j = 0; j < 3; j++) {
         scanf("%d", &amat[i][j]);
      }
   }

   printf("\nEnter Second Matrix:n");
   for (i = 0; i < 3; i++) {
      for (j = 0; j < 3; j++) {
         scanf("%d", &bmat[i][j]);
      }
   }

   printf("The First Matrix is: \n");
   for (i = 0; i < 3; i++) {
      for (j = 0; j < 3; j++) {
         printf(" %d ", amat[i][j]);
      }
      printf("\n");
   }

   printf("The Second Matrix is : \n");
   for (i = 0; i < 3; i++) {
      for (j = 0; j < 3; j++) {
         printf(" %d ", bmat[i][j]);
      }
      printf("\n");
   }

   //Multiplication Logic
   for (i = 0; i <= 2; i++) {
      for (j = 0; j <= 2; j++) {
         sum1 = 0;
         for (k = 0; k <= 2; k++) {
            sum1 = sum1 + amat[i][k] * bmat[k][j];
         }
         cmat[i][j] = sum1;
      }
   }

   printf("\n Multiplication Of Two Matrices : \n");
   for (i = 0; i < 3; i++) {
      for (j = 0; j < 3; j++) {
         printf(" %d ", cmat[i][j]);
      }
      printf("\n");
   }

   return (0);
}
```

Result

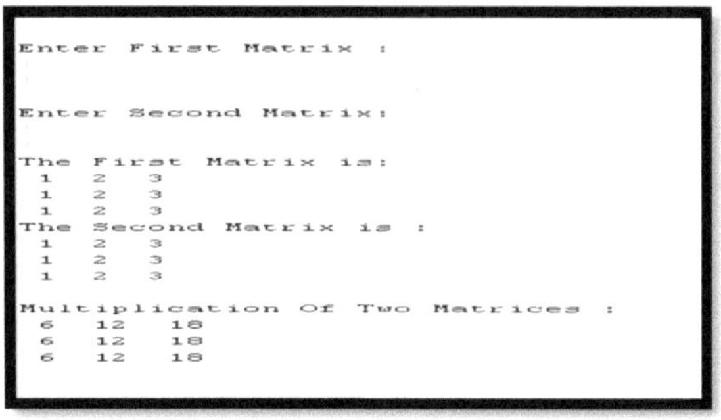

```
Enter First Matrix :


Enter Second Matrix:

The First Matrix is:
 1   2   3
 1   2   3
 1   2   3
The Second Matrix is :
 1   2   3
 1   2   3
 1   2   3

Multiplication Of Two Matrices :
 6   12   18
 6   12   18
 6   12   18
```

- **Transpose of Matrix**

```
#include <stdio.h>

void main()
{
    static int arrayt[10][10];
    int i, j, k, l;

    printf("Enter the order of the matrix \n");
    scanf("%d %d", &k, &l);
    printf("Enter the coefiicients of the matrix\n");
    for (i = 0; i < k; ++i)
    {
        for (j = 0; j < l; ++j)
        {
            scanf("%d", &arrayt[i][j]);
        }
    }
    printf("The given matrix is \n");
    for (i = 0; i < k; ++i)
    {
        for (j = 0; j < l; ++j)
        {
            printf(" %d", arrayt[i][j]);
        }
        printf("\n");
    }
    printf("Transpose of matrix is \n");
    for (j = 0; j < l; ++j)
    {
        for (i = 0; i < k; ++i)
        {
            printf(" %d", arrayt[i][j]);
        }
        printf("\n");
    }
}
```

Result

```
Enter the order of the matrix

Enter the coefiicients of the matrix


The given matrix is
 2 4
 5 7
Transpose of matrix is
 2 5
 4 7
```

- Inverse of Matrix

```
#include<stdio.h>
#include<math.h>

float determinant1(float[][],float);
void cofactor1(float[][],float);
void transpose1(float[][],float[][],float);
int main()
{
  float a1[10][10],k,d;
  int i,j;
  printf("------------------------------------------------------------\n");
  printf("----------------made by Dr.J.S.Parab ------------------------\n");
  printf("------------------------------------------------------------\n");
  printf("\n  C Program to find inverse of Matrix\n\n");
  printf("Enter the order of the Matrix : ");
  scanf("%f",&k);
  printf("Enter the elements of %.0fX%.0f Matrix : \n",k,k);
  for (i=0;i<k;i++)
    {
     for (j=0;j<k;j++)
       {
        scanf("%f",&a1[i][j]);
        }
    }
  d=determinant1(a,k);
  printf("Determinant of the Matrix = %f",d);
  if (d==0)
   printf("\n Matrix Inverse not possible\n");
  else
   cofactor1(a1,k);
   printf("\n\n**** Thank for using the program!!! ****");
```

```
}

/*For calculating Determinant of the Matrix */
float determinant1(float a[10][10],float k)
{
  float s=1,det=0,b[25][25];
  int i,j,m,n,c;
  if (k==1)
    {
     return (a1[0][0]);
    }
  else
    {
     det=0;
     for (c=0;c<k;c++)
       {
        m=0;
        n=0;
        for (i=0;i<k;i++)
          {
            for (j=0;j<k;j++)
              {
                b[i][j]=0;
                if (i != 0 && j != c)
                 {
                   b[m][n]=a[i][j];
                   if (n<(k-2))
                    n++;
                   else
                    {
                     n=0;
                     m++;
                     }
                   }
                }
              }
          det=det + s * (a1[0][c] * determinant1(b,k-1));
          s=-1 * s;
          }
    }

    return (det);
}

void cofactor1(float num[25][25],float f)
{
 float b[25][25],fac[25][25];
 int p,q,m,n,i,j;
 for (q=0;q<f;q++)
 {
   for (p=0;p<f;p++)
    {
     m=0;
     n=0;
     for (i=0;i<f;i++)
     {
       for (j=0;j<f;j++)
        {
          if (i != q && j != p)
          {
            b[m][n]=num[i][j];
            if (n<(f-2))
             n++;
            else
             {
```

```
                n=0;
                m++;
                }
             }
          }
        }
        fac[q][p]=pow(-1,q + p) * determinant1(b,f-1);
      }
    }
    Transpose1(num,fac,f);
}
/*Finding transpose of matrix*/
void transpose1(float num[25][25],float fac[25][25],float r)
{
  int i,j;
  float b[25][25],inverse[25][25],d;

  for (i=0;i<r;i++)
    {
     for (j=0;j<r;j++)
       {
        b[i][j]=fac[j][i];
       }
    }
  d=determinant1(num,r);
  for (i=0;i<r;i++)
    {
     for (j=0;j<r;j++)
       {
       Inverse1[i][j]=b[i][j] / d;
       }
    }
   printf("\n\n\nThe inverse of matrix is : \n");

   for (i=0;i<r;i++)
    {
     for (j=0;j<r;j++)
       {
        printf("\t%f",inverse1[i][j]);
       }
    printf("\n");
     }
}
```

Result

```
---------------------------------------------------------------------
----------------made by Dr.J.S.Parab --------------------------
---------------------------------------------------------------------

  C Program to find inverse of Matrix

Enter the order of the Matrix :
Enter the elements of 2X2 Matrix :

Determinant of the Matrix = 2.000000

The inverse of matrix is :
        2.000000          -1.000000
        -1.500000         1.000000

**** Thanks for using the program!!! ****
```

6.8 Reading from the Flash (Web Application)

A Web server is a platform that stores the content such as Web pages and delivers to the clients as and when requested. Here, we have implemented FPGA DE2 development board based on Web server. Web server core is first instantiated in a Nios II system. Nios II system for Web application is designed using SOPC Builder of Quartus II CAD. After implementing Web server Nios II system, an C++ application program can be run on to system to implement the Web server. Here, the Web pages are loaded beforehand on to the flash memory of the DE2 board.

Procedure

- Open the Quartus software and create a new project.
- Go to assignments, select import assignments, and add the de2_pin assignment file.
- Select create tcl file for project from the project menu and the run the tcl script by selecting tcl script from the tool menu.
- To select the components, open the SOPC Builder and choose the following components (Fig. 6.9)
  - NIOS II PROCESSOR (STANDARD)
  - JTAG UART
  - SDRAM
  - SRAM

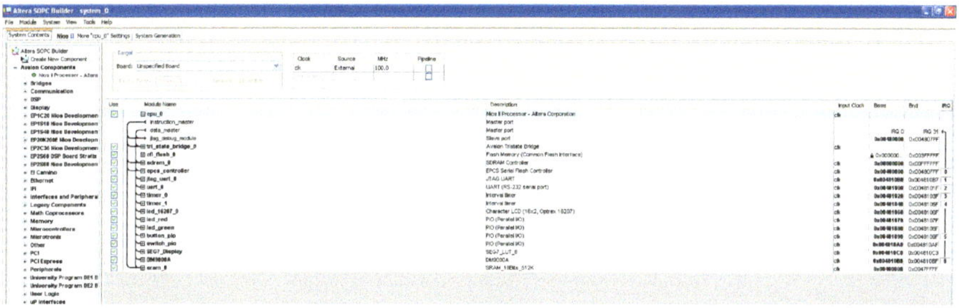

Fig. 6.9 SOPC components selected

- Flash
- Timer
- PIO
- LEDs
- LCD.

- Next auto-assign base addresses and Irq.
- In the system generation tab, click generate.
- Move on back to the Quartus software and click on file menu and choose new block and schematic file.
- Right click on the workspace and add the component created in SOPC (located in the project folder).
- Add the respective connectors to the I/O generated for the component.
- Save the entire project with the same file name as the entity and compile.
- In tools, choose programmer, check the program, configure tab, and click start.
- Open the Nios II IDE software.
- In the file menu, choose new C/C++ program, and from the template, choose zip file system project.
- Do the necessary changes like setting the base address in the system library.
- Build the entire program and run on hardware.

Note: before compiling the project, make sure you have loaded the files in the flash using the DE2 control panel utility. The files should be ziped in .zip format with 0% compression.

Steps for Programming Web Pages on to FLASH Memory

- The flash has to be programmed with the Web pages of the Web server.
- Program the FPGA with the usb_api.sof.
- Execute the DE2_control_panel.exe and switch to the flash page as shown below.

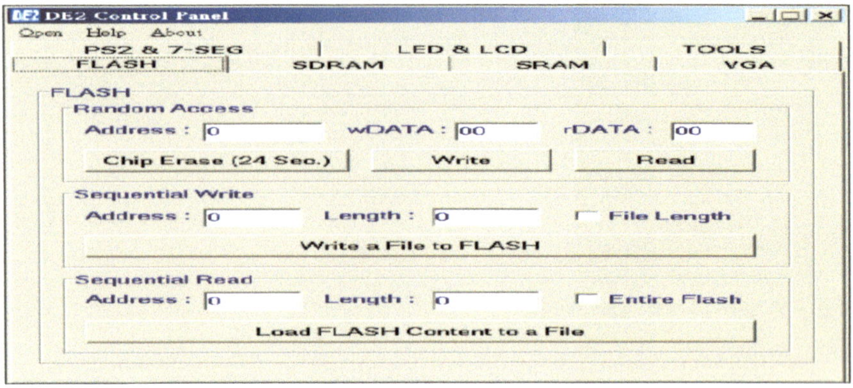

- Click the "Chip Erase" button to erase flash memory.
- Select the "File length" checkbox and confirm that start address is 0 under sequential Write.
- Click the "write a file to flash" button. Select the ro_zipfs.zip in your local directory (DE2_web).
- Close DE2_panel.
- Open the Quartus software and open the de2_web server project.
- Launch the Nios II IDE and open the DE2_web workspace.
- Build and compile the C code given below.
- Run the design choosing "run" as hardware.
- Open the Web browser and input the IP into the address bar which is displayed on the LCD.
- *Static ip address: 192. 168. Sw[15:12]. Sw[11:8]+128.*
- *Mac address: 00-90-00-ae-sw[7:0].*
- By changing the position on switches as mentioned above, the static IP and mac address can be changed.

Full System for Web Application

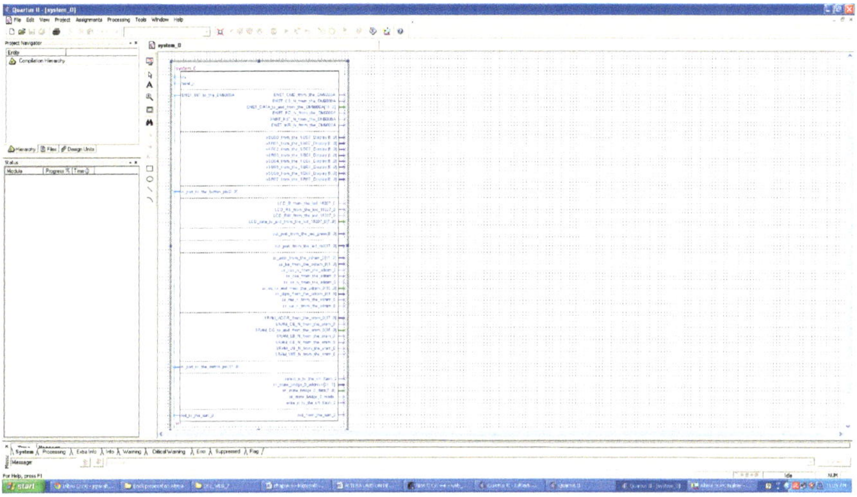

Compilation Report

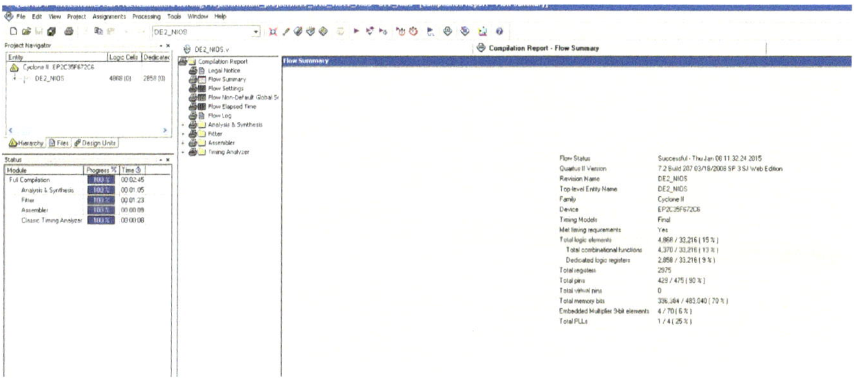

```
----------------------C code for web server Implementattion-----------------------
#include <stdio.h>
#include <errno.h>
#include <ctype.h>
#include "includes.h"
#include "alt_lwip_dev.h"
#include "lwip/sys.h"
#include "user.h"


#include "dm9000.h"
#include "lcd.h"
ALTERA_AVALON_DM9K_INSTANCE(DM9000A, dm9k);
```

```
void user_task(void * pvoid)
{
    static u_long val=0;

        for(;;)
    {
        val ^= (1<<17);
        IOWR(LED_RED_BASE, 0, val);
        usleep(500000);
    }
}

#ifndef LWIP
  #error This Web Server requires the Lightweight IP Software Component.
#endif

#ifndef __ucosii__
  #error This Web Server requires the UCOS II IP Software Component.
#endif

#ifndef  RO_ZIPFS
  #error This Web Server requires the Altera Read only Zip filing system.
#endif

OS_EVENT *attained_ip_address_sem;

static void tcpip_init_done(void *arg)
{

    ALTERA_AVALON_DM9K_INIT(dm9k);

 if (!lwip_devices_init(ETHER_PRIO))
   die_with_error("[tcpip_init_done] Fatal: Can't add ethernet interface!");

   attained_ip_address_sem = OSSemCreate(1);

#if LWIP_DHCP == 1
    if(!(IORD(SWITCH_PIO_BASE, 0) & (1<<17)))
sys_thread_new(dhcp_timeout_task, NULL, DHCP_TMR_PRIO);

#endif

  if(!sys_thread_new(http_task, NULL, HTTP_PRIO))
    die_with_error("[tcpip_init_done] Fatal: Can't add HTTP task!");

}

int main ()
{
    LCD_Init();
   lwip_stack_init(TCPIP_PRIO, tcpip_init_done, 0);
    sys_thread_new(user_task, NULL, SANITY_PRIO);

  OSStart();

  return 0;
}
```